STRUCTURAL PACKAGING ART

WITH DVD-ROM

ARTPOWER

Structural Packaging Art

Designer: Chen Ting
Chief Editor: Xia Jiajia

Address: Suites 2001, 20/F., Chinachem Tower, 36 Connaught Road
Central, Hong Kong, China
Tel: 852 3184 0676
Fax: 852 2543 2396

Editorial Department
Address: G009, Floor 7th, Yimao Centre, Meiyuan Road, Luohu
District, Shenzhen, China
Tel: 86-755-82913355
Fax: 86-755-82020029

URL: www.artpower.com.cn
E-mail: artpower@artpower.com.cn

ISBN 978-988-14687-3-4

Printed in China

PREFACE

Despite the fact that we are living in a very diverse and rapidly changing world, packaging design will not undergo global changes in near future.

Different forces affect packaging design development: changes in communication structure, retail development, including the emergence of completely new formats, as well as change of the environment, especially visual culture and aesthetics.

If looking a little deeper, it seems that the most important changes that likely affect the packaging will be in the consumer's behavior and existential sphere.

All of these changes look so global and significant that colorful futuristic forecasts involuntary appear in our minds.

However... Looking back in time, based on my experience in package design (almost 15 years), there wasn't many crucial changes.

The first fridge with internet connection was launched in 2000. I was probably not the only one who thought that in a couple of years our refrigerators will order food by themselves. So we could "dig a grave" for the packaging design, because the "cold" fridge mind needs no design, only barcode.

Eco-friendly packaging. It is still a beautiful slogan. If "eco pack" manages to win serious positions, it can happen only if "eco" goes hand in hand with some other functions: usability, aesthetics or price. The good old paper still dominates among the eco-friendly materials, but now it can be used in a different way: with the "craft" color codes, "natural" claims, minimalistic printing.

With changes in communication structure, i.e. move from mass media to Internet and social networks, there was expectations, that new consumer, who creates its own information field and own content, will be particularly susceptible to the package personalization.

Despite this and the fact that new technology makes the personalization opportunities more accessible and interesting day by day, this approach to design has not become widespread, although looks good at design festivals.

There is a lot of work that can be done to improve functional features of the goods such as: easy to use, to store and to recycle.

How often do you see people in the store, scanning the packaging's QR-code?

Those changes, which are really worth to pay attention to understand the most up-to-date packaging solutions, take place in the existential sphere. Initially, oversupply and high competition level made the powerful impulse to branding and packaging design and it ended with the appearance of "consumer society". Now it is expected that the part of the society, which feel themselves "over-consumed" already, will be different in choosing products and packages. We already see this: the "terrible" word commoditization that describes the phenomenon, which is the result of this trend. Nowadays, not every product is so passionately desired as yesterday.

Packaging development for such products is a huge scope for innovations, which will be easier and more interesting with technology up-grading. There is much work can be done to improve functional features of the goods such as: easy to use, to store, to recycle. Of course, there will be some adaptation to the new retail opportunities and requirements.

So what will change in the package of brands, which will manage to keep top place in consumer's minds? In a world where everything is changing so rapidly, copied quickly and one trend replaces another, the importance of consistency and authenticity will definitely increase. And if we talk about packaging design in particular within branding story, the values of solution will be determined not by the degree of novelty, but by the building of individuality in long time perspective.

Kugaevskikh Andrey, Creative Director, Svoe Mnenie

CONTENTS

HOUSEHOLD PACKAGE

FOOD PACKAGE

BEVERAGE PACKAGE

MISCELLANEOUS

HOUSEHOLD PACKAGE

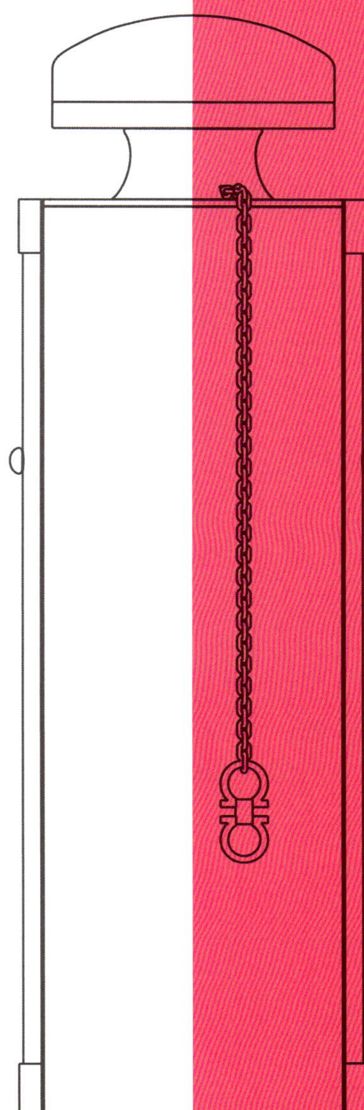

THE MINI FRAGRANCES COLLECTION

Designed by Sotano Studio, the Mini Fragrance Collection is the delightful new line by Salvatore Ferragamo. Each little spinning top style bottle encloses 20ml of our favourite Signorina and Incanto fragrances.

¨The project transversal to all Ferragamo feminine perfumes had to capture the essence and the character of each fragrance in one volume design.

We wanted to create little jewels people would love to collect and a smooth and pleasant shape easy to keep in a purse¨. Founders Fanny Le Bonniec & Oscar Del Cerro.

Inspired by the whirling spin tops of our childhood, the little sprays swiftly travel the world just like all Ferragamo creations. A playful expression o f creativity that blends tradition and innovation, it takes Italian class to a whole new level.

Design Agency:
Sotano Studio

Designer:
Fanny Le Bonniec & Oscar Del Cerro

Client:
Salvatore Ferragamo

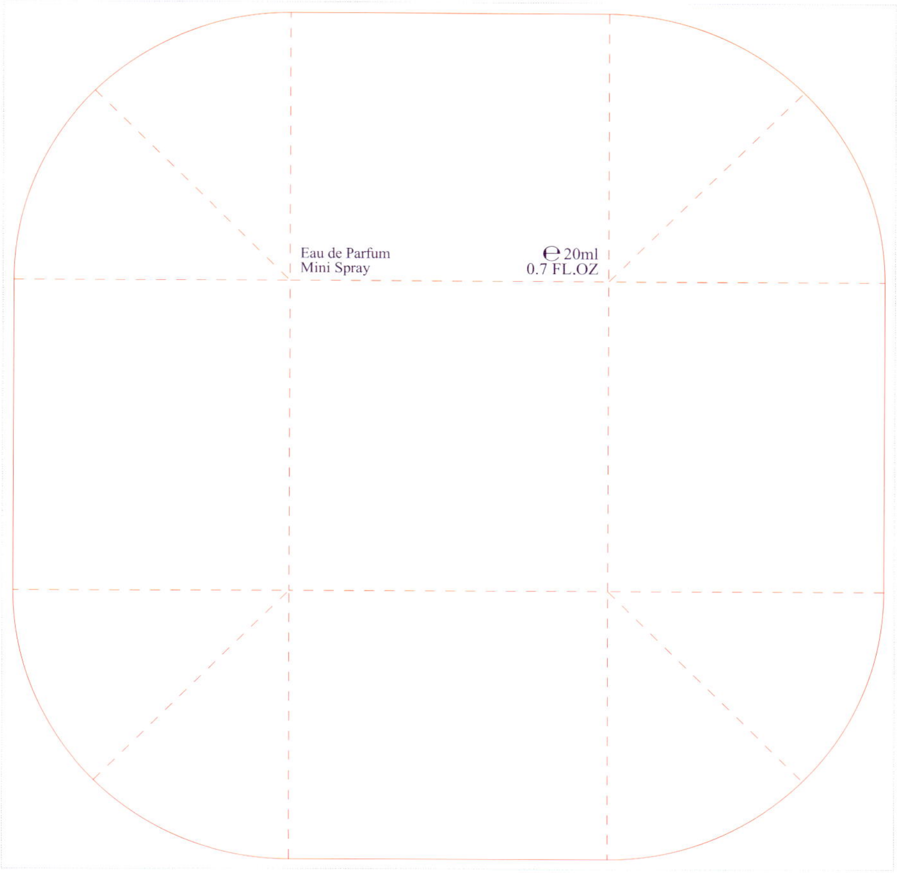
Eau de Parfum
Mini Spray
℮ 20ml
0.7 FL.OZ

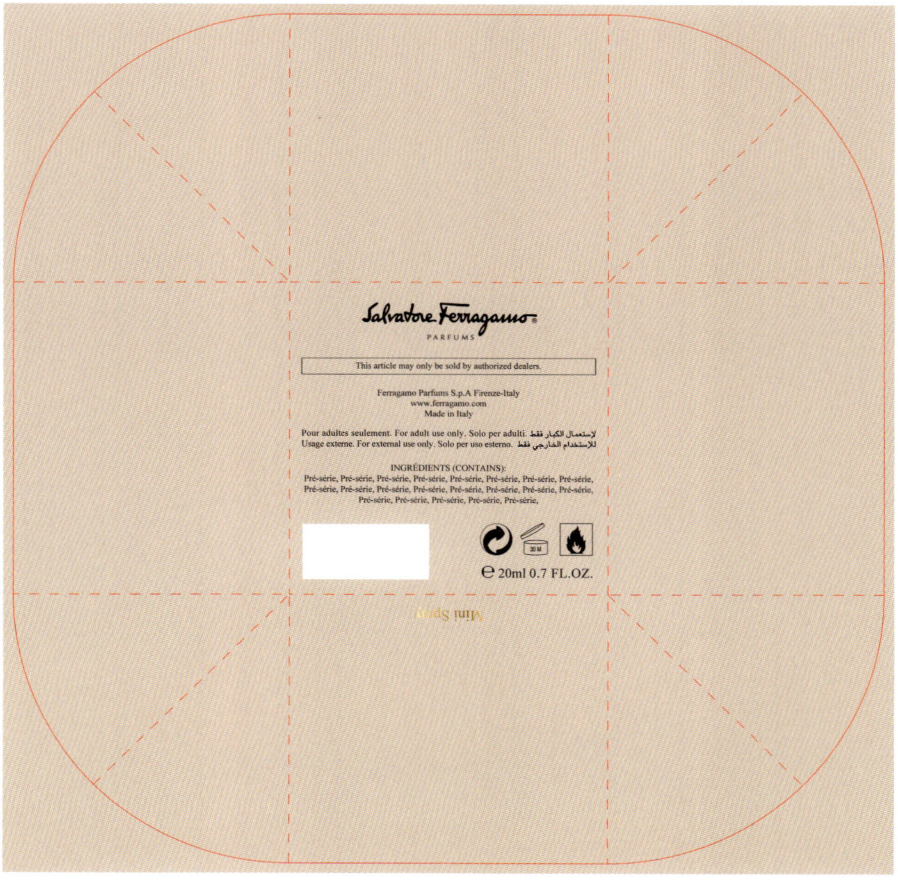
Salvatore Ferragamo
PARFUMS
This article may only be sold by authorized dealers.
Ferragamo Parfums S.p.A Firenze-Italy
www.ferragamo.com
Made in Italy
Pour adultes seulement. For adult use only. Solo per adulti. لإستعمال الكبار فقط
Usage externe. For external use only. Solo per uso esterno. للإستخدام الخارجي فقط
INGRÉDIENTS (CONTAINS):
Pré-série, Pré-série, Pré-série, Pré-série, Pré-série, Pré-série, Pré-série, Pré-série,
Pré-série, Pré-série, Pré-série, Pré-série, Pré-série, Pré-série, Pré-série, Pré-série,
Pré-série, Pré-série, Pré-série, Pré-série, Pré-série,
30 M
℮ 20ml 0.7 FL.OZ.
Mini Spray

The precious outer box is half-rigid half-foldable and opens like a surprise box. Inside the bottle lying on a soft foam pad is ready to start "making its style spins across the world" as stated by the quote and spin-tops golden icons.

The merry-go-round glorifier, also designed by the agency, continues the storytelling and to trigger the latent child in our adulthood.

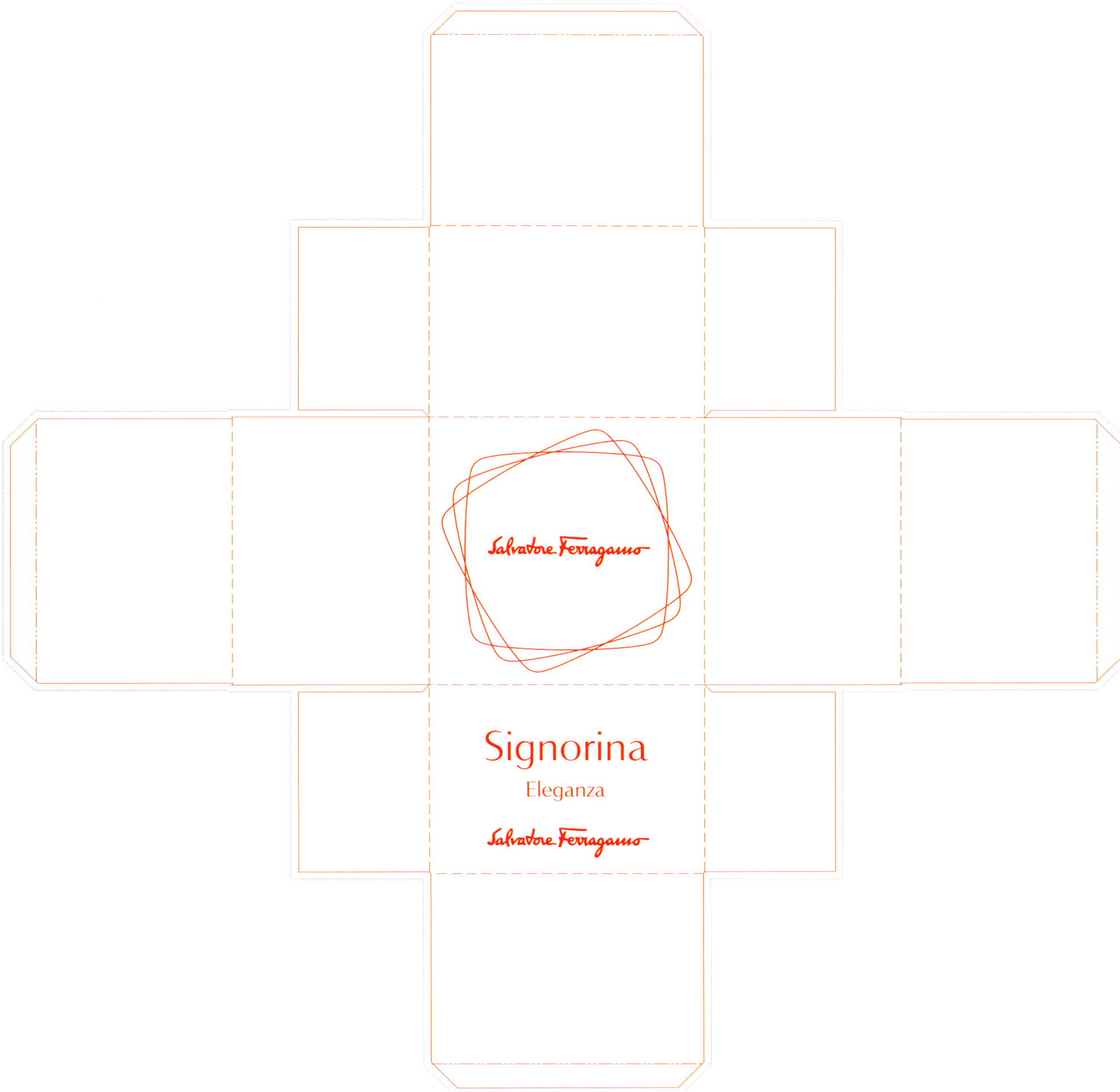
Salvatore Ferragamo
Signorina
Eleganza
Salvatore Ferragamo

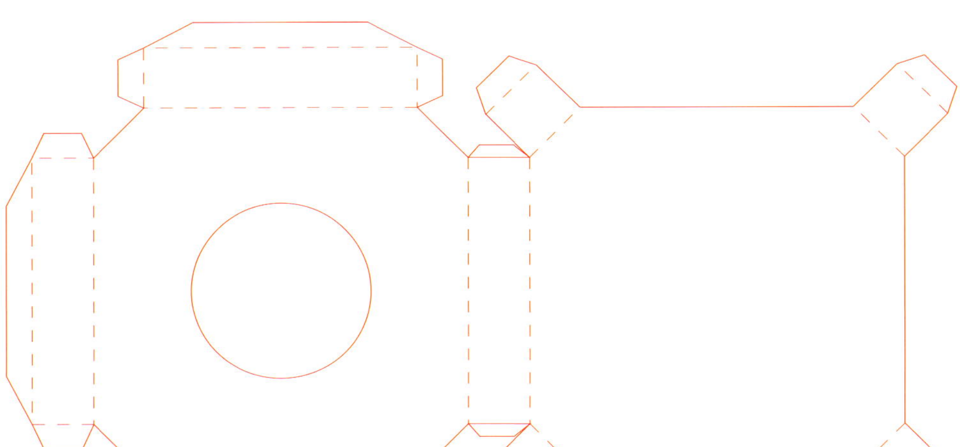

SILHOUETTE BOTTLE & PACKAGING

Silhouette bridges the gap between two worlds; being soft and romantic, but also being powerful through structure and shape. The bottle, inspired by the bright and crisp fabrics designer Christian Siriano often incorporates into his designs, signifies how the fragrance itself unfolds in ethereal layers.

Design Agency:
Karim Rashid

Designer:
Karim Rashid

Client:
Christian Siriano/ The Fragrance Group

Photography:
Karim Rashid

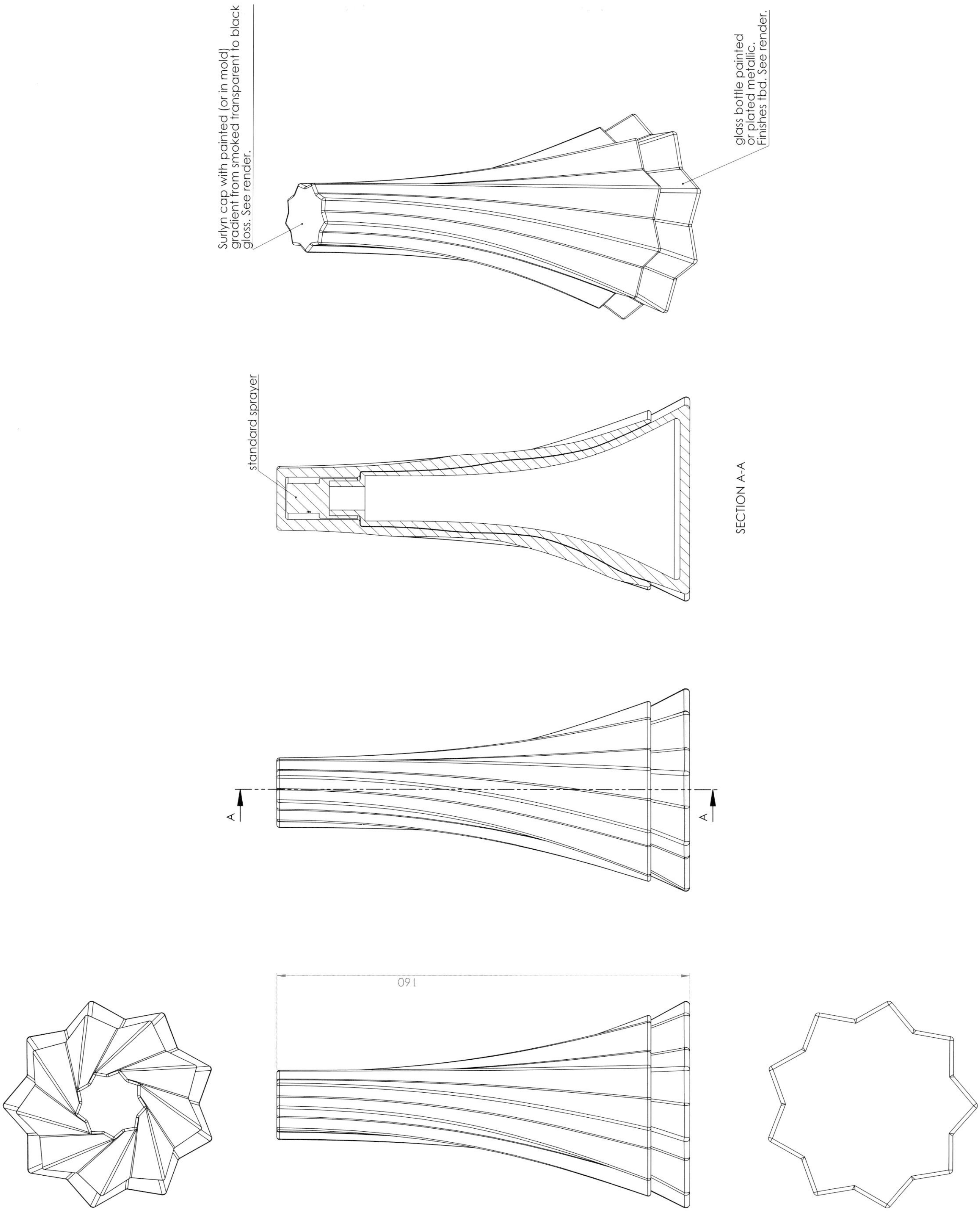
Surlyn cap with painted (or in mold) gradient from smoked transparent to black gloss. See render.
glass bottle painted or plated metallic. Finishes tbd. See render.
standard sprayer
SECTION A-A
A
A
160

KENZO JEU D' AMOUR BOTTLE

Fresh, sensual, floral, and fruity, the fragrance is inspired by the senses of love and euphoria. The fragrance is composed of notes of tea, floral accords of tuberose and freesia, pomegranate, mandarin, and a sensual base of sandalwood, and musk. The 50 ml glass fragrance bottle is enchanting with a sensual, confident, straight-lined silhouette accented by the lush pink hue of the perfume, making it warm and inviting. The exquisite packaging, combined with the lively bouquet of the fragrance itself, embodies Jeu D'Amour's passionate dedication to the thrills of romance.

Design Agency:
Karim Rashid

Designer:
Karim Rashid

Client:
Kenzo

Photography:
Karim Rashid

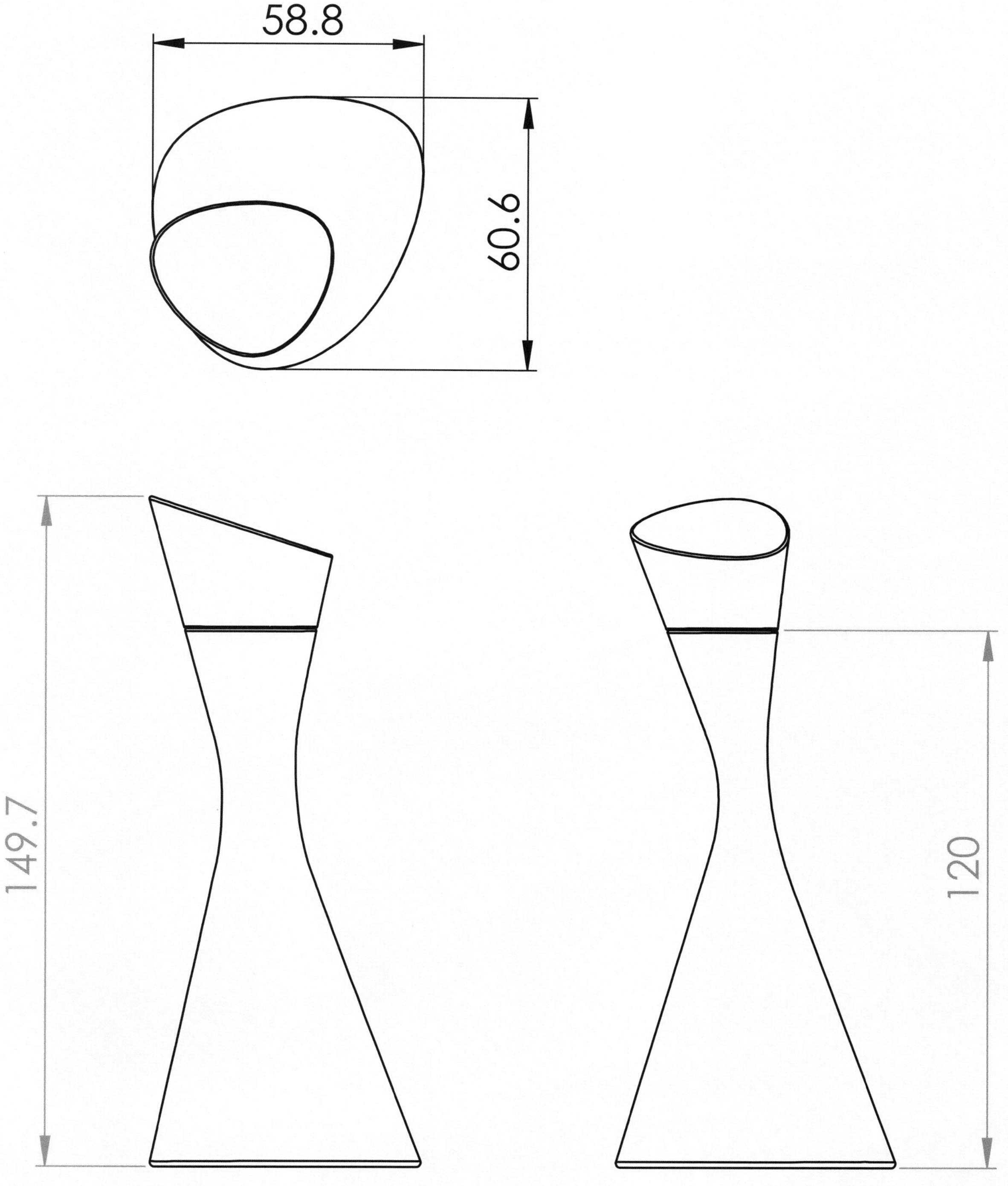
58.8
60.6
149.7
120

Design Agency:
ESTABLISHED

Photography:
Stephanie Dinkel

CALVIN KLEIN REVEAL

Bottle design for Calvin Klein fragrance: Reveal. The bottle is a blend of the strict, minimal, geometric proportions synonymous with the Calvin Klein brand and femininity expressed through pillow-like curves. The cap has a high polished finish in soft gold, complimenting the warmth of the juice color and cool of the glass.

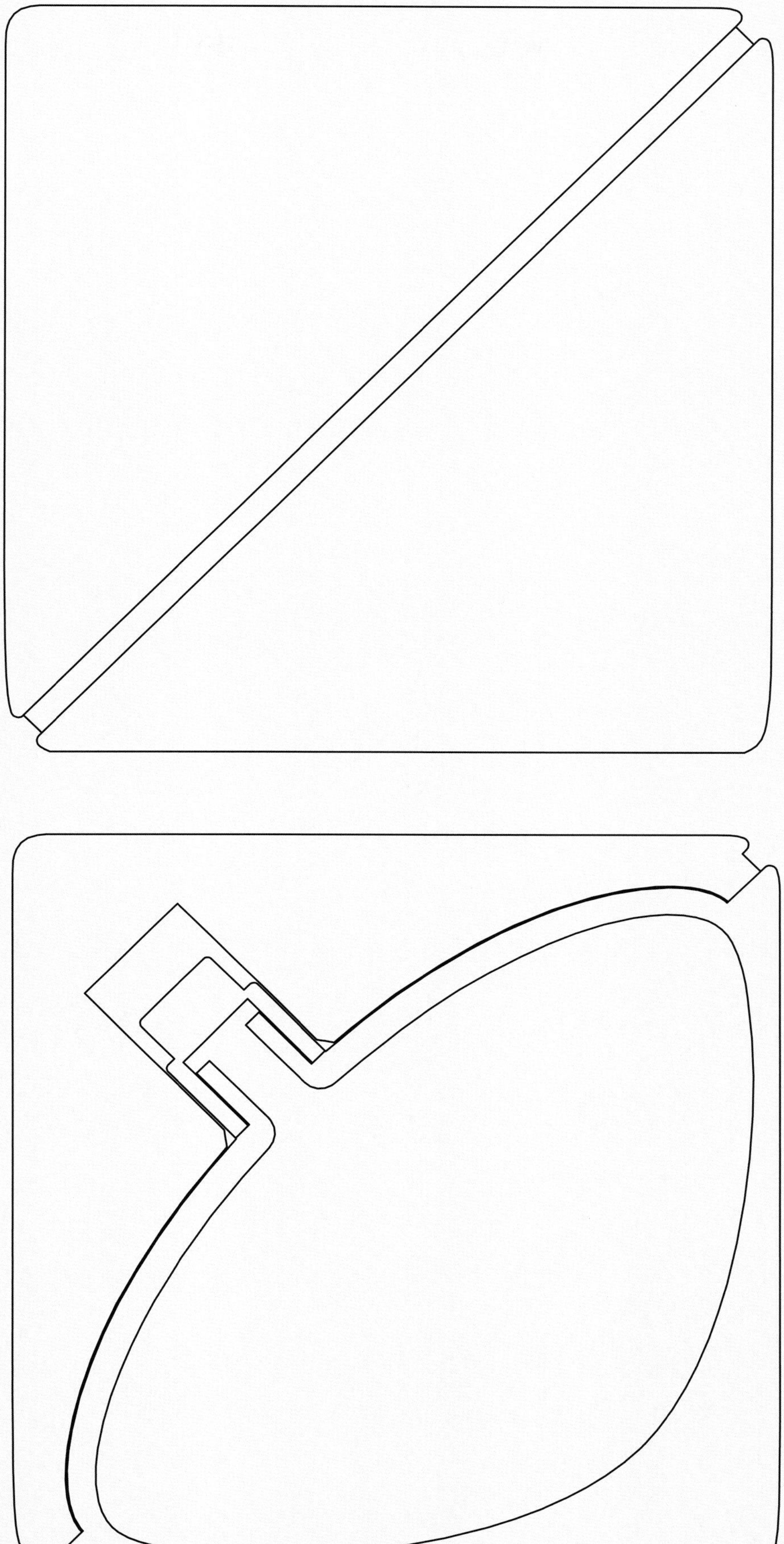

VICTORIA'S SECRET - ANGEL'S ONLY, HEAVENLY, FEARLESS

Established developed three new fragrance bottles for Victoria's Secret.

Design Agency:
ESTABLISHED

Photography:
Stephanie Dinkel

Angels Only

VICTORIA'S SECRET
FEARLESS

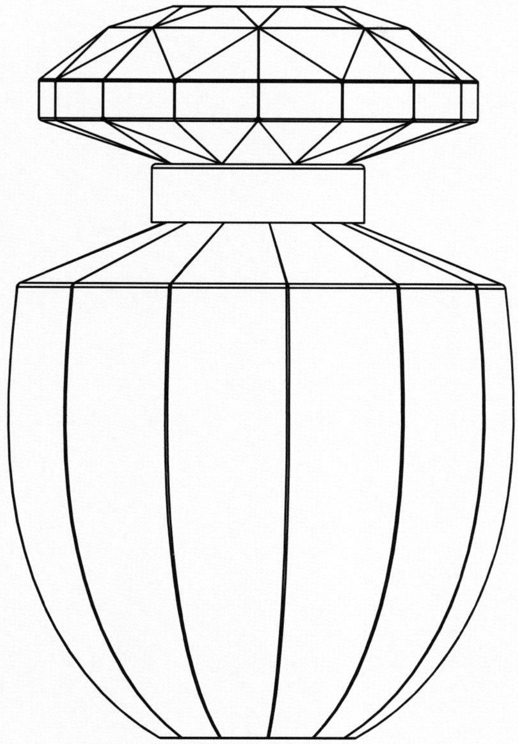

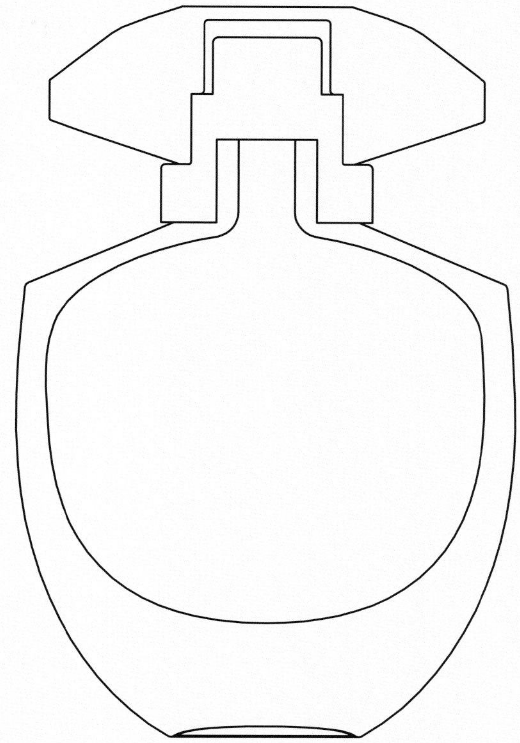

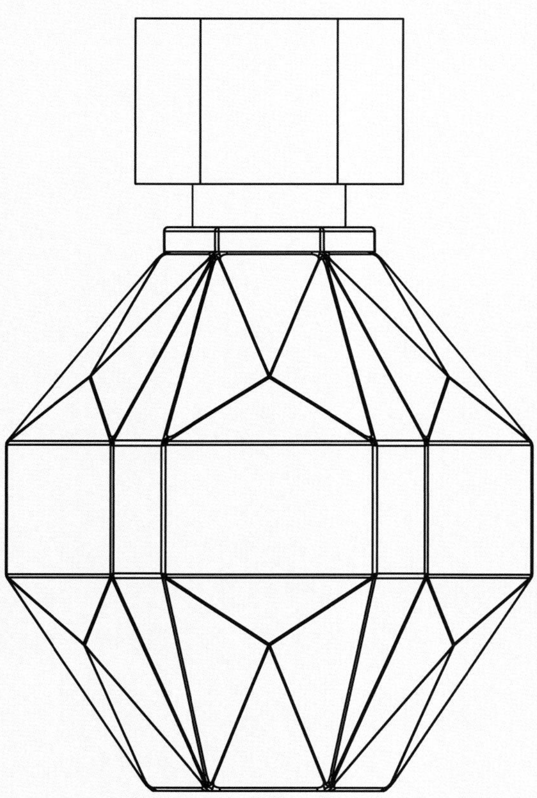

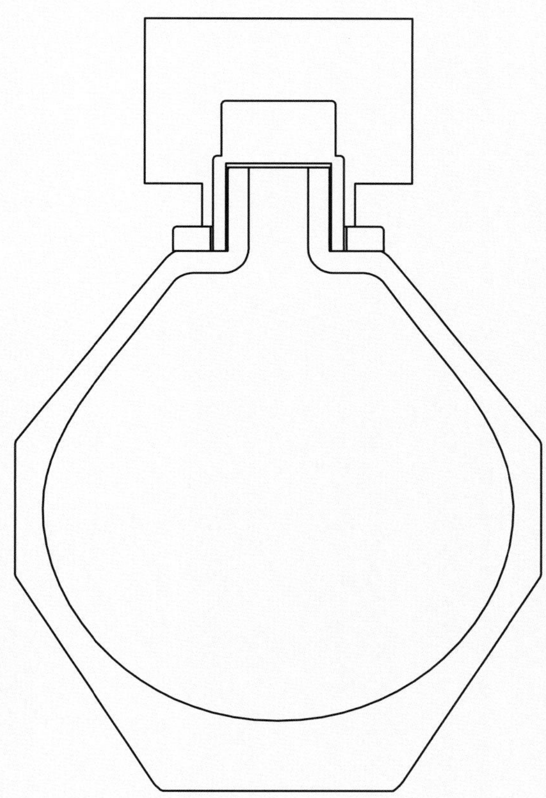

KIA EAU DE TOILETTE

The auto maker, which used to be considered serious-minded, designed the signature perfume for use by men and women of all ages based on the sensory marketing.

The contemporary and sophisticated perfume was designed which incorporates the brand attributes such as "vibrant", "reliable", and "distinctive".

The three underlying key words of "vibrant", "reliable", and "distinctive" were applied to the perfume design suited to the brand strategy that excludes the decorative elements and revolves around the design concept of minimalism accentuated by modernity. The identity was expressed by dynamic line on cold metallic material to highlight the fact that this perfume was made by auto maker.

The capsule-like containers and cap that have the same shape were designed to ensure overall volume balance. They are accentuated by the modern and simple body with sophisticated wave line of contradictory material. The silver matt metal and dark gray color added sophistication and elegance.

After the glass blow molding, the metals are manually assembled. The matt aluminum body improved the grip during the use and provides the connecting part to the container and cap, increasing the directional convenience during the capping.

The cost reduction and mass production can be achieved by the automated production.

Design Agency:
Aekyung Design Center

Creative Director:
Chang Kum Suk

Designer: Jung Young Soo / Hong Sung Eun

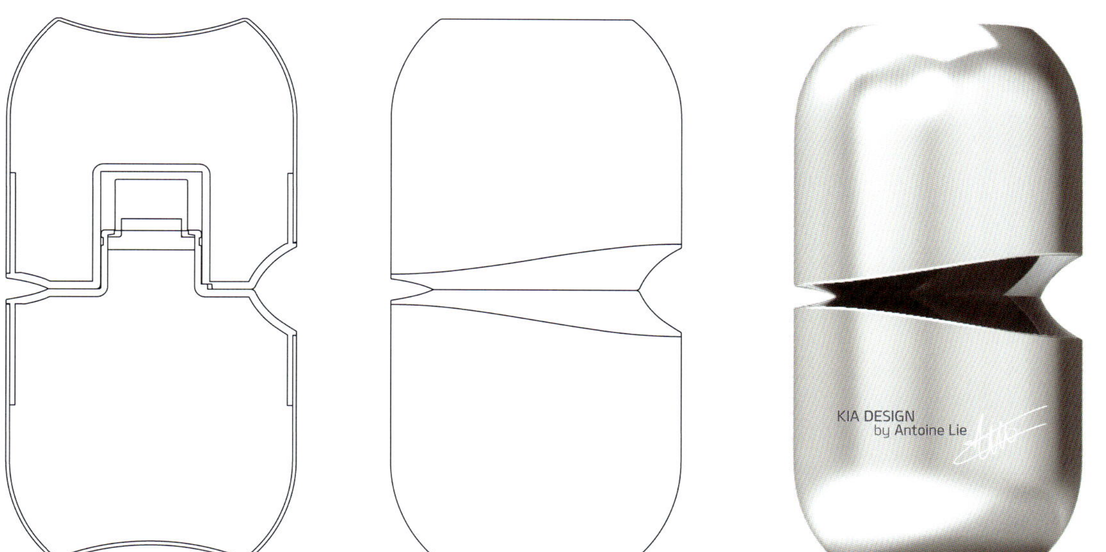
KIA DESIGN
by Antoine Lie

L'EAUNDRY - FRAGRANCE LAUNDRY DETERGENT

Treat your second skin the way your skin should be treated.

Concept and a design for a new laundry detergent by The Deli Garage: L'EAUNDRY, a luxury laundry detergent that smells and looks like a high-class perfume. To treat your second skin like your first. Bottles in the shape of vintage chemists' bottles establish the connection to washing agents while the typography and formal style fully invoke the world of fine scents. Available in two scents: Figue pour femme. Olibanum pour homme.

Design Agency:
KOREFE. Kolle Rebbe Form und Entwicklung.

Creative Director/Head of Design:
Christian Doering (Concept)

Designer:
Christian Doering, Katharina Ullrich

Client:
T.D.G. Vertriebs GmbH & Co. KG

Photography:
Katharina Ullrich

Copywriter:
Lorenz Ritter

Product Management:
Shari Linthe

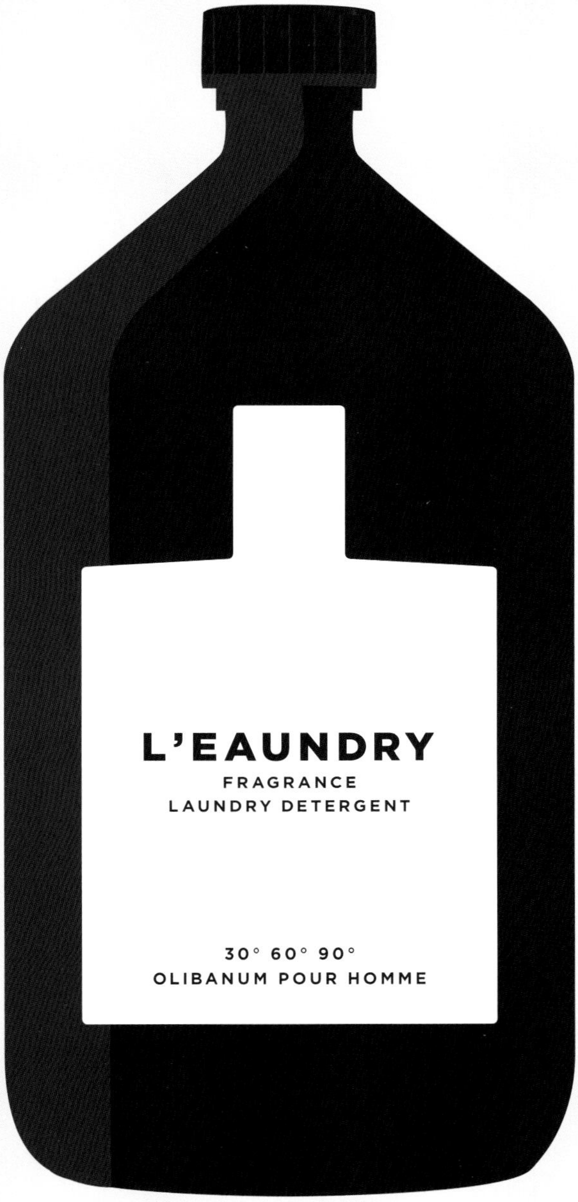
L'EAUNDRY
FRAGRANCE
LAUNDRY DETERGENT
30° 60° 90°
OLIBANUM POUR HOMME

L'EAUNDRY
FRAGRANCE
LAUNDRY DETERGENT
30° 60° 90°
FIGUE POUR FEMME

MARC JACOBS DAISY DREAM

Design of fragrance bottle for Marc Jacobs Daisy Dream.

In a fresh and new approach, the iconic daisy was used to make a type of shawl around the "shoulders" of the bottle reflecting the playful femininity of the brand.

Design Agency:
ESTABLISHED

Photography:
Stephanie Dinkel

EMOZIONE

Emozione, a multifaceted fragrance that combines feminine floral bouquets to a refined and precious woody texture enriched with a touch of sensual, glamorous chypre notes.

Designer:
PATRICK VEILLET

Client:
FERRAGAMO

Copyright :
Patrick Veillet

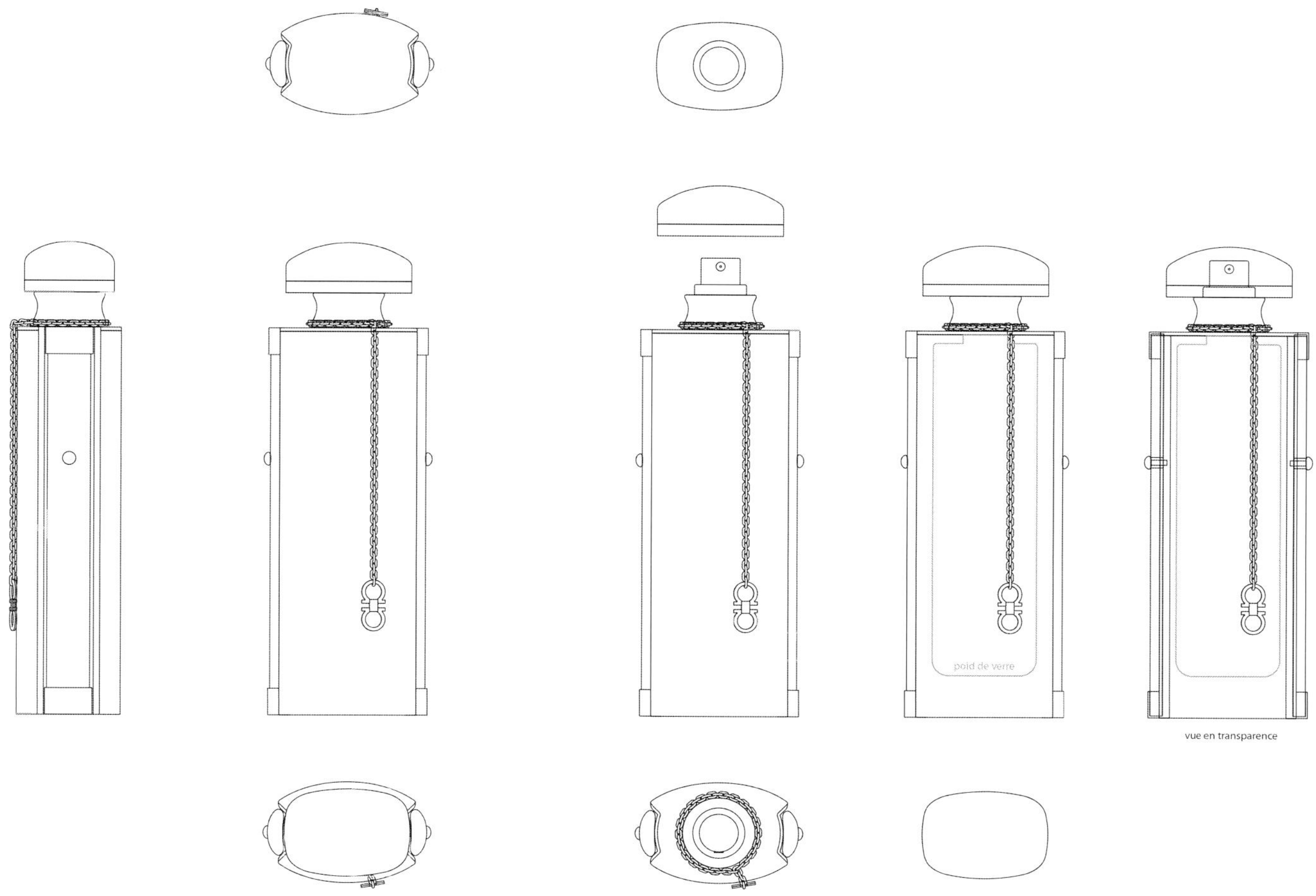
poid de verre
vue en transparence

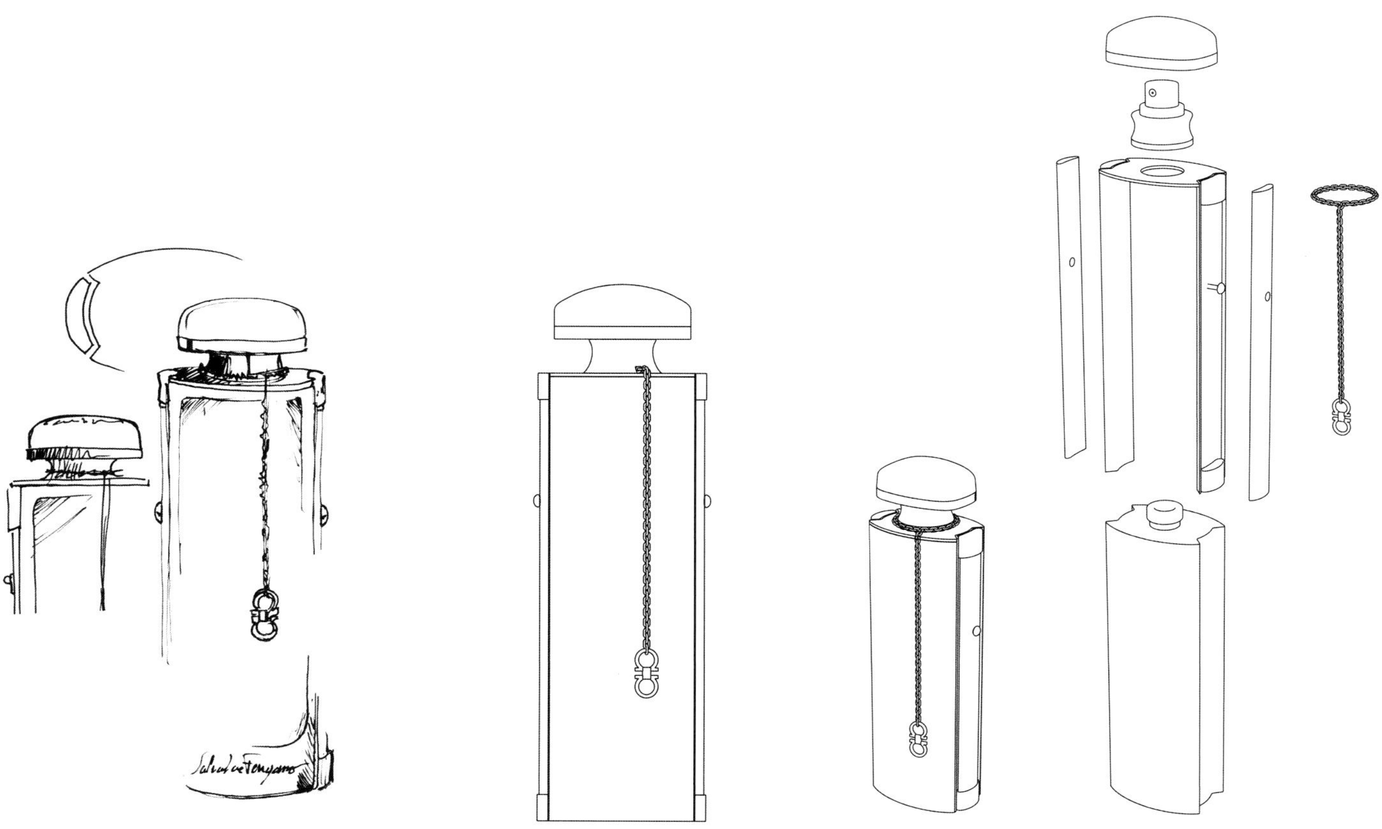

BROOKS BROTHERS BLACK FLEECE

This fragrance packaging for Brooks Brothers sought to combine the traditional world of fine tailoring with the distinctly forward thinking attitude of Thom Browne's Black Fleece collection. Ceramics and references to office stationary of the 1940s were reinterpreted in a modern way. A zip-strip on the secondary packaging tears off to remove all graphics, revealing a pure white inner box, containing layers of cut-out felt used in suit-making to reflect the heritage of the brand in tailoring. The packaging is 100% recyclable.

Design Agency:
ESTABLISHED

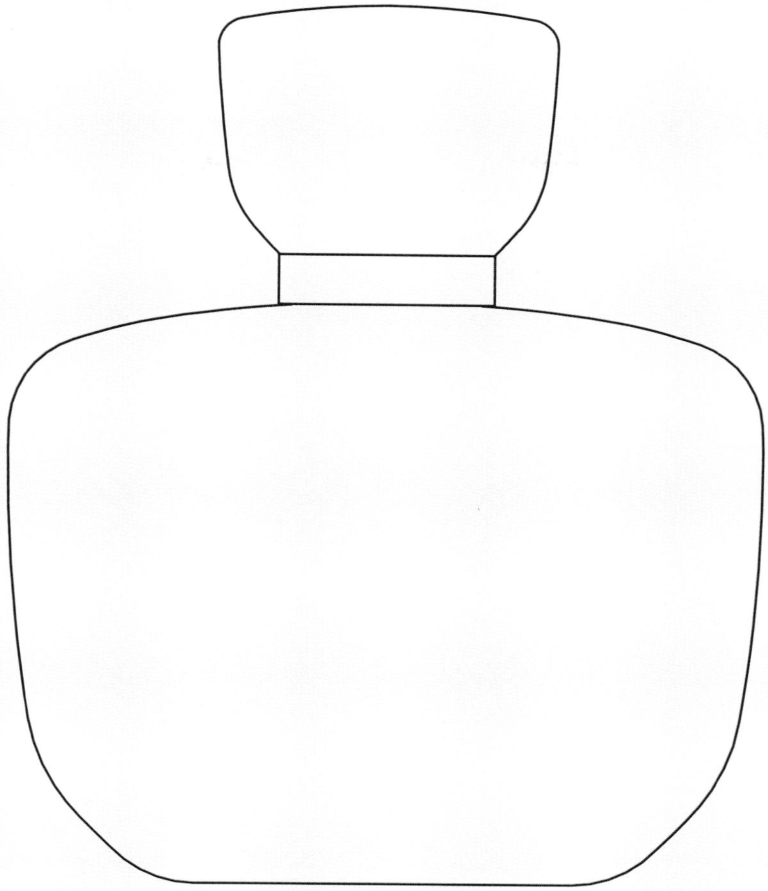

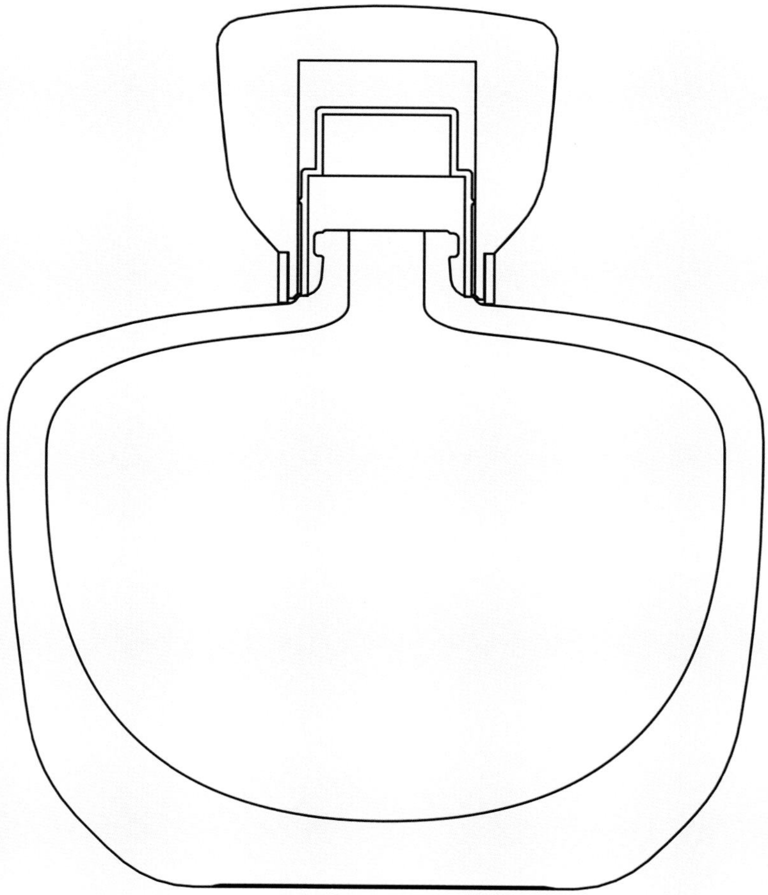

CATCH ME

By taking on this project, I wanted above all to express delicious pleasure, surprise and lightness in a bottle. Little by little, the playfulness, movement and elegance of a jewel enriched this story. The result is a singular and seductive object that's incredibly clever. The harmony of contrasts is also a key element of this project.

On the one hand, we have this somewhat "twisted" glass that recalls waves on calm water for "liquid" rigidity. On the other, we have these metallic bubbles of different sizes, both hard and soft. They bring to mind an imposing jewelry piece for women, but also the marbles from our childhood. All of these elements come together in a random movement, defined by the hand that plays with them. A multi-faceted and multi-sensory object that you discover anew each time you pick it up. An addictive object that stirs up an irresistible desire to grab it!

The Catch me... woman is a natural, spontaneous and coquettish young woman with a good sense of humor that makes her sparkle. Her smile is her first weapon of seduction. She dares to play. She proudly accepts this need for playfulness in her life as a young woman.

Designer:
PATRICK VEILLET

Client:
CACHAREL

Copyright :
Patrick Veillet

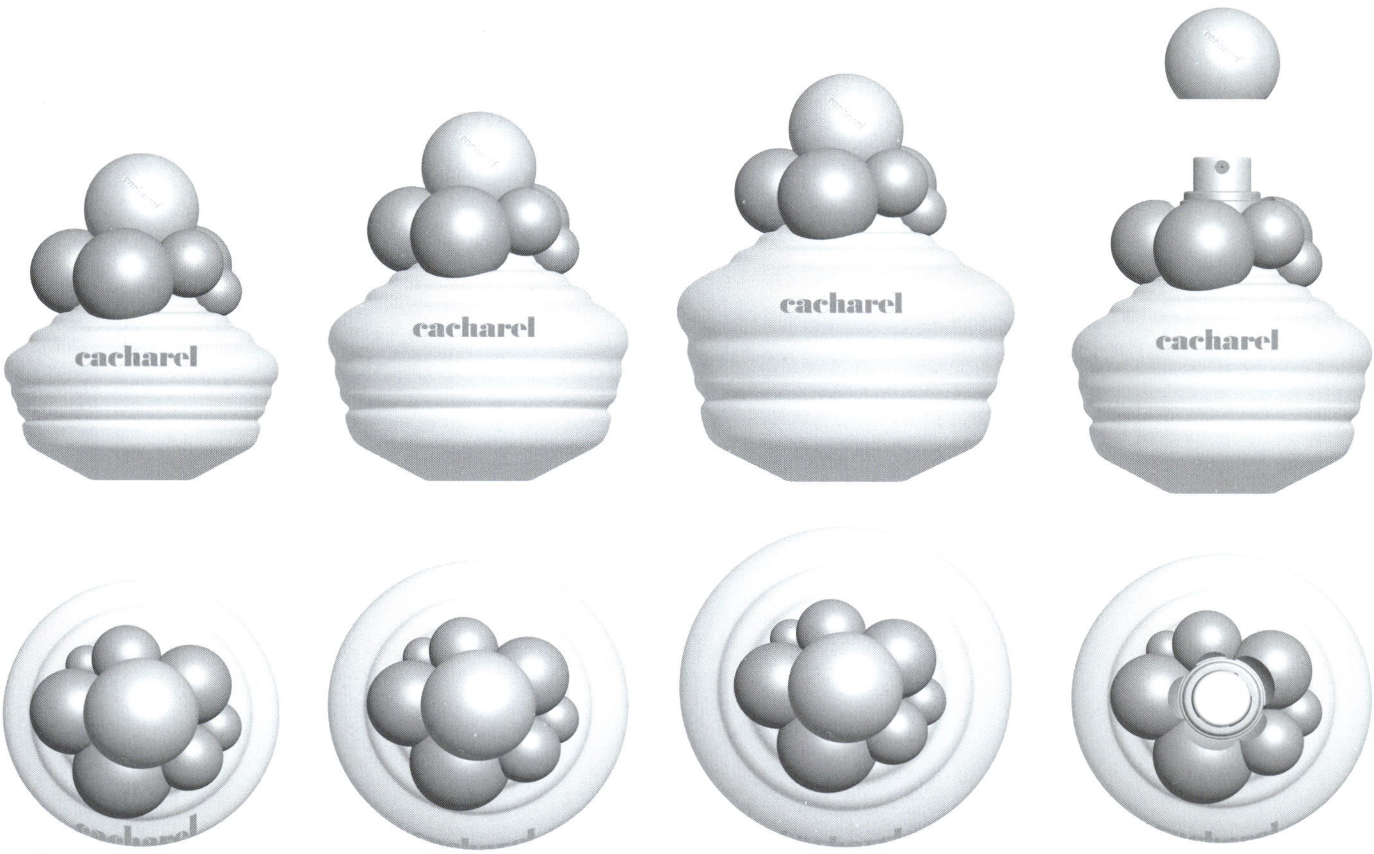

30ml 50ml 80ml 50ml OPEN

N-CIGALE

It is after working five years in the « Haute-Couture » industry, having signed since 2001 symbolic perfume bottles and advised the leading fashion, luxury and beauty marks that Patrick VEILLET decides in 2010 to establish his own mark.

EMBLEM of APOLLON since the Antique Greece, symbol of the BEAUTY, the ARTS and the CREATION, the fascinating CICADA becomes the source of his inspiration which will lead him to create N-CIGALE.

"This Goddess of the Dawn", worshipped from immemorial times in Asia by the Emperors to North America by the Indians, is a sacred symbol of protection and a promise of eternity.

To restore the respectability to this Ancestral and Unique Symbol of Art and Beauty, Patrick VEILLET imagines a singular Collection of jewelry and objects of decoration for the house by appealing for their realization to art craftsmen renowned for their excellence as well as to the sharpest technologies.

Designer:
PATRICK VEILLET

Client:
N-CIGALE

Copyright :
Patrick Veillet

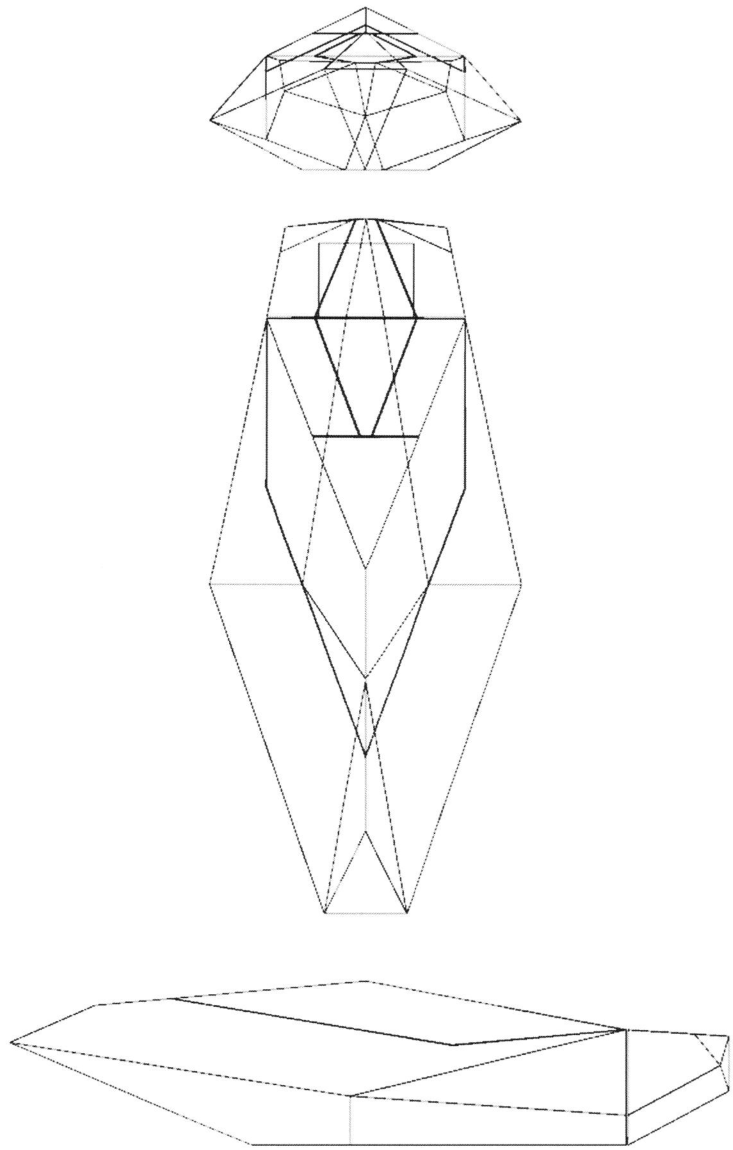

N-CIGALE
LES CHANTS PARFUMÉS

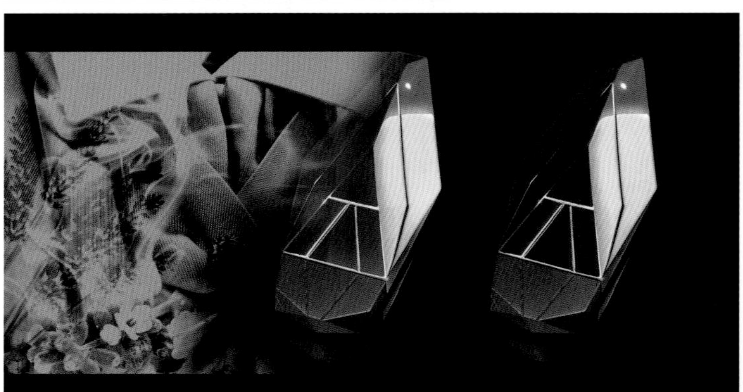

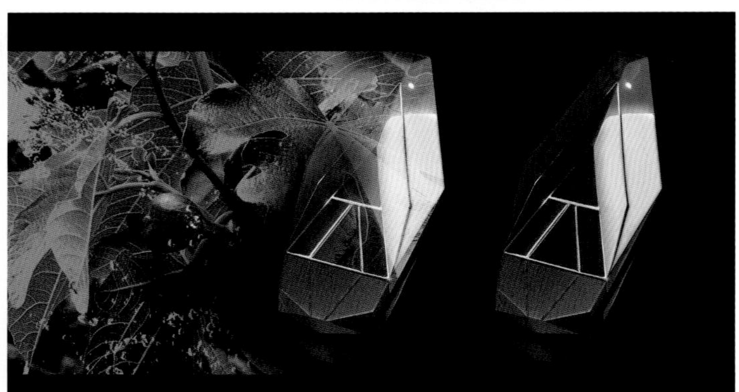

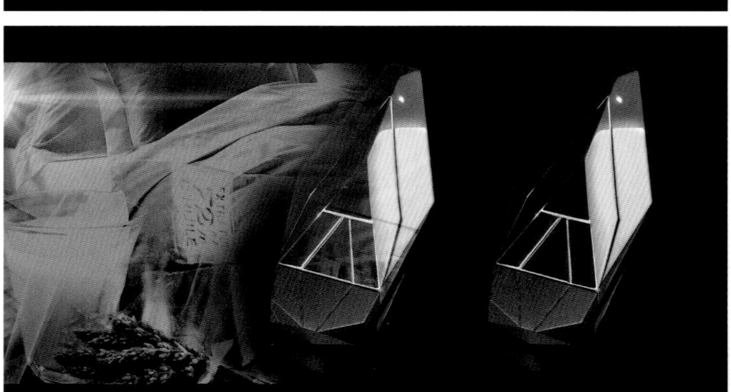

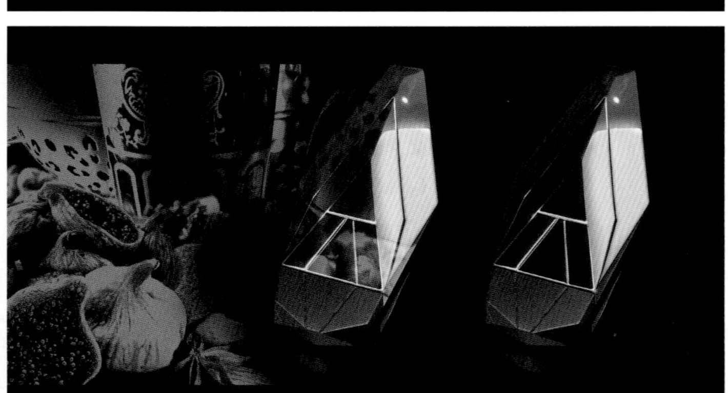

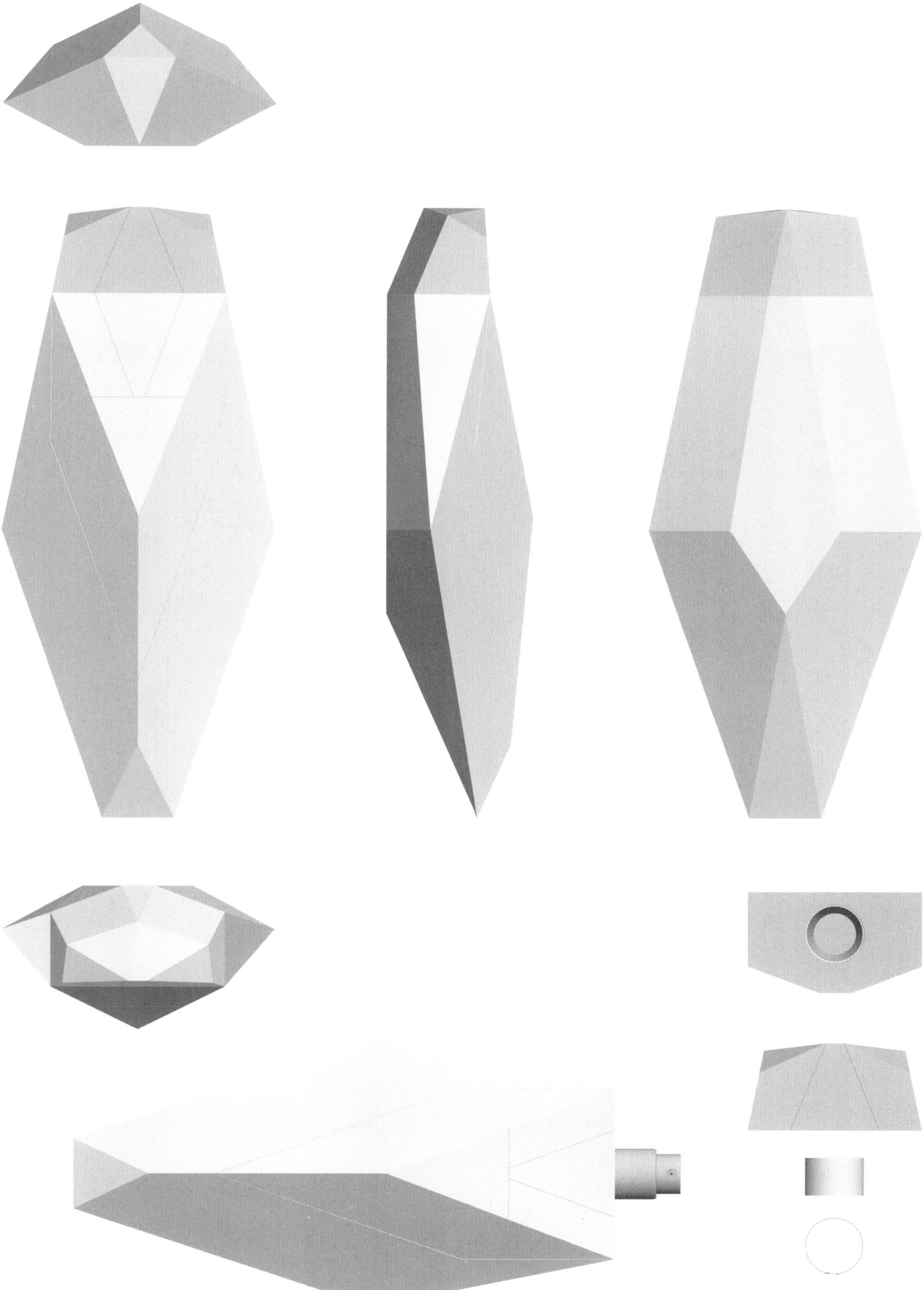

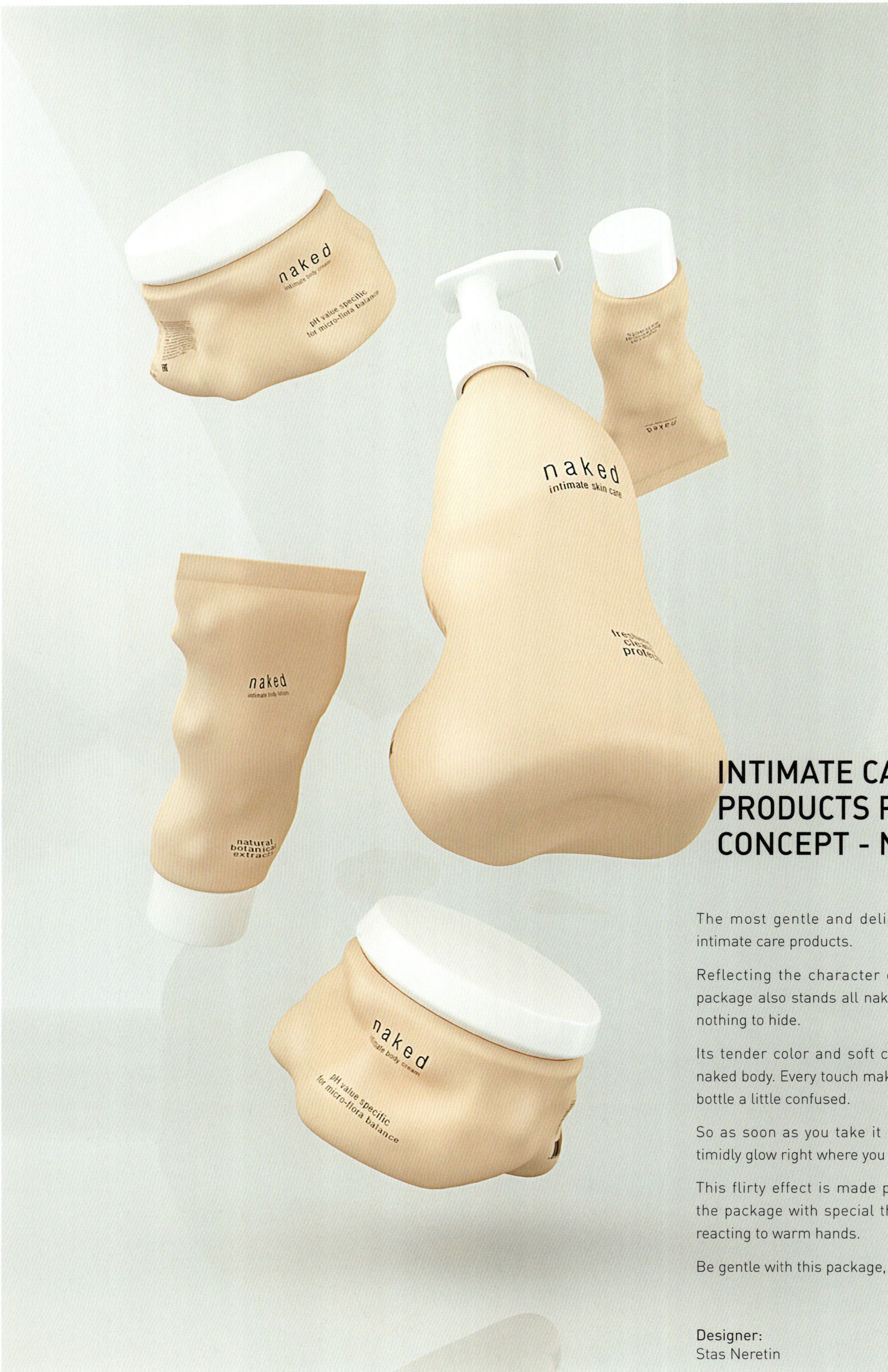

INTIMATE CARE PRODUCTS PACKAGE CONCEPT - NAKED

The most gentle and delicate cosmetics are intimate care products.

Reflecting the character of its contents, the package also stands all naked before you having nothing to hide.

Its tender color and soft curves resemble the naked body. Every touch makes a tube, a cup or a bottle a little confused.

So as soon as you take it in your hand, it will timidly glow right where you touch it.

This flirty effect is made possible by covering the package with special thermochromic paint reacting to warm hands.

Be gentle with this package, it is very shy.

Designer:
Stas Neretin

NAKED
NAKED

naked
intimate body lotion
naked
intimate skin care
freshens
cleans
protects
naked
intimate body lotion
natural
botanical
extracts
naked
intimate body cream
pH value specific
for micro-flora balance

naked
intimate body cream
pH value specific
for micro-flora balance
250 ml

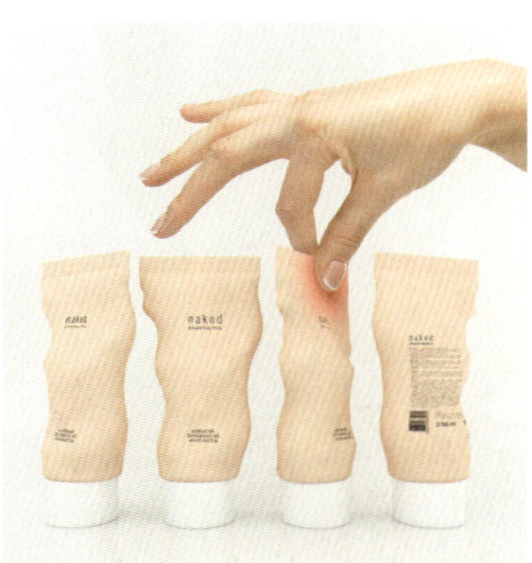

naked
freshens
cleans
protects

naked

NAKED
NAKED
NAKED

VILDE

This was a design project initiated by the Graphic Design course at Malmö University, Sweden. The assignment involved developing a concept for a perfume, come up with a name, packaging and how to market it. The perfume was to be linked to your choice of brand and target group.

VILDE is a unisex fragrance created from natural ingredients. The name VILDE means "Savage" in Swedish. It represents the natural ingredients of the product, as well as something about the person using it.

My choice of brand was Filippa K, a Scandinavian fashion brand. I liked the idea of having perfumes for different seasons—hence the Spring/Summer and Autumn/Winter editions. It also gave the perfumes a "fashion" touch, and an opportunity to match them with the brand's clothing collections.

The target group for Filippa K is modern women and men who like clean, minimalist clothes and subtle scents. High-end without being too luxurious.

Designer:
Linda Hurtig

VILDE Perfume Box 4.5 x 12.2 x 4.5 cm

CHA TZU TANG LAUNDRY DETERGENT

Worries and distress are sometimes inevitable with our daily responsibilities and challenges. It is only when we are back to our cozy home that we feel safe and secured. From laundry, dishes to fruits and vegetables, we provide you with the purest gift from nature, building a prefect home as it deserves to be. Our very fine, non-toxic foam is absolutely gentle on your hands and harmless to the environment.

Design Agency:
Victor Branding Design Crop.

Client:
Cha Tzu Tang

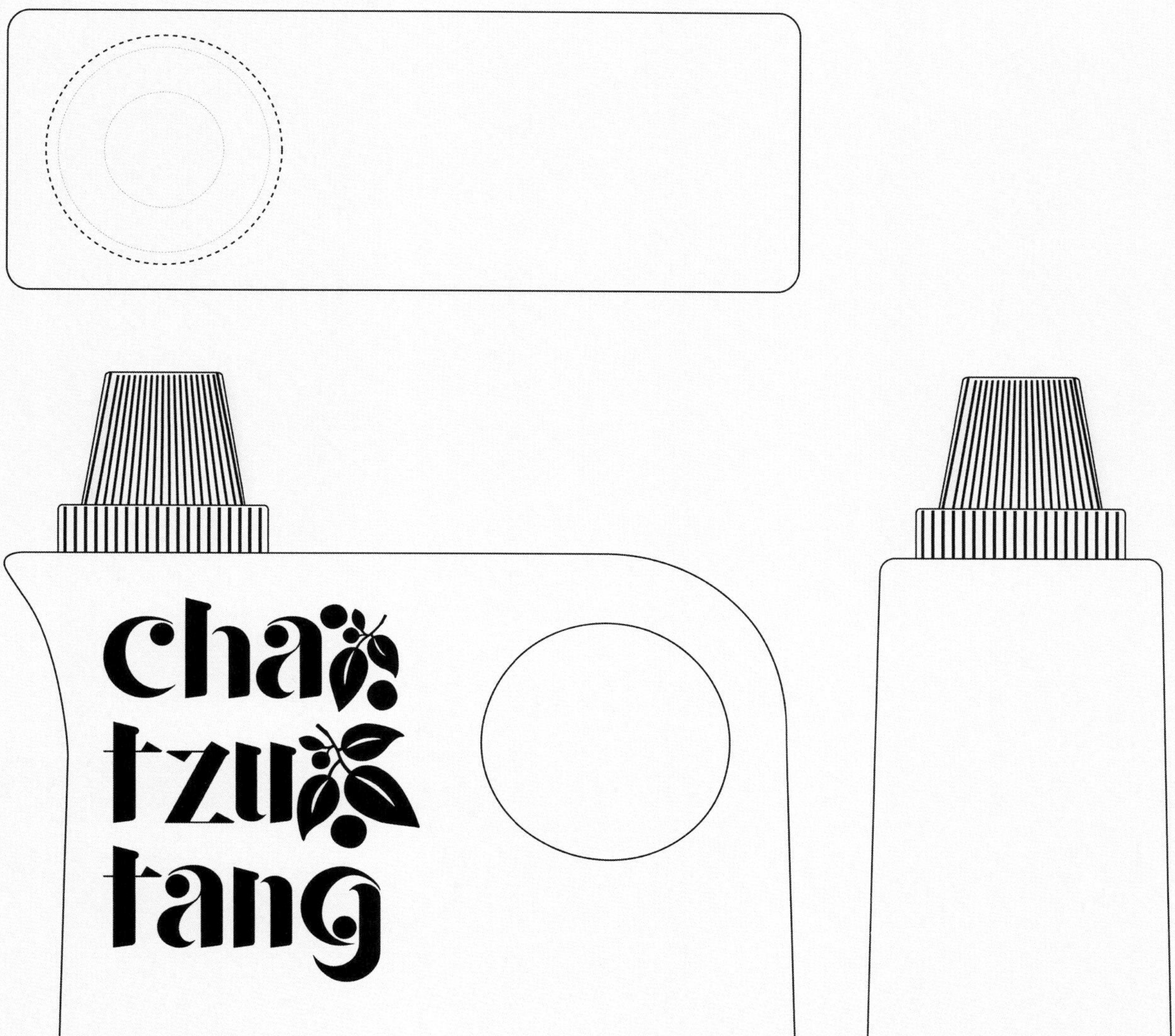
cha
tzu
tang

LOVE AT FIRST SIGHT

Ansell is a global leader in sexual wellness and protection. Their range of personal lubricant needed to distinguish itself from the competition and increase stand out on shelf in a way that did not negate the sensual nature of these products or cause embarrassment to customers in-store. The new packaging by Reverse Innovation should also arouse curiosity and convey a sense of trust.

Fun and sensuality are the core creative ideas which drive this highly distinctive and engaging redesign. The iconic silhouette has become synonymous with the brand's identity and signifies a truly innovative development in the visual language of the product category. The upper ring is irresistible and encourages interaction. The delivery mechanism is highly ergonomic and a delight to use – gel flow is instant and can be easily controlled while the ring handle gives the product its distinctive playful character.

The brand's key benefits are communicated in a clear, uncluttered graphic style featuring blushing tonal gradients to suggest both the rising sensation of pleasure and fragrance variant. The product coveys an image that is freewheeling and accessible which appeals to regular and new consumers alike. By shaking up the category, Ansell has been empowered with an ownable and distinctive brand identity.

Design Agency:
Reverse Innovation

Creative Director:
Alice Tacconi

Art Director:
Mirco Onesti

Structural Designer:
Gustavo de Souza Messias

Project Manager:
Peter Solomon

Illustrator:
Valerio Tizzi

Graphic Designer:
Luca Mazzoleni

Client:
Ansell Ltd.

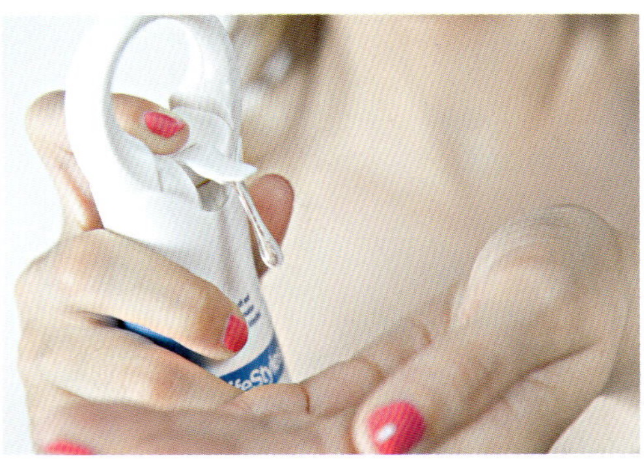

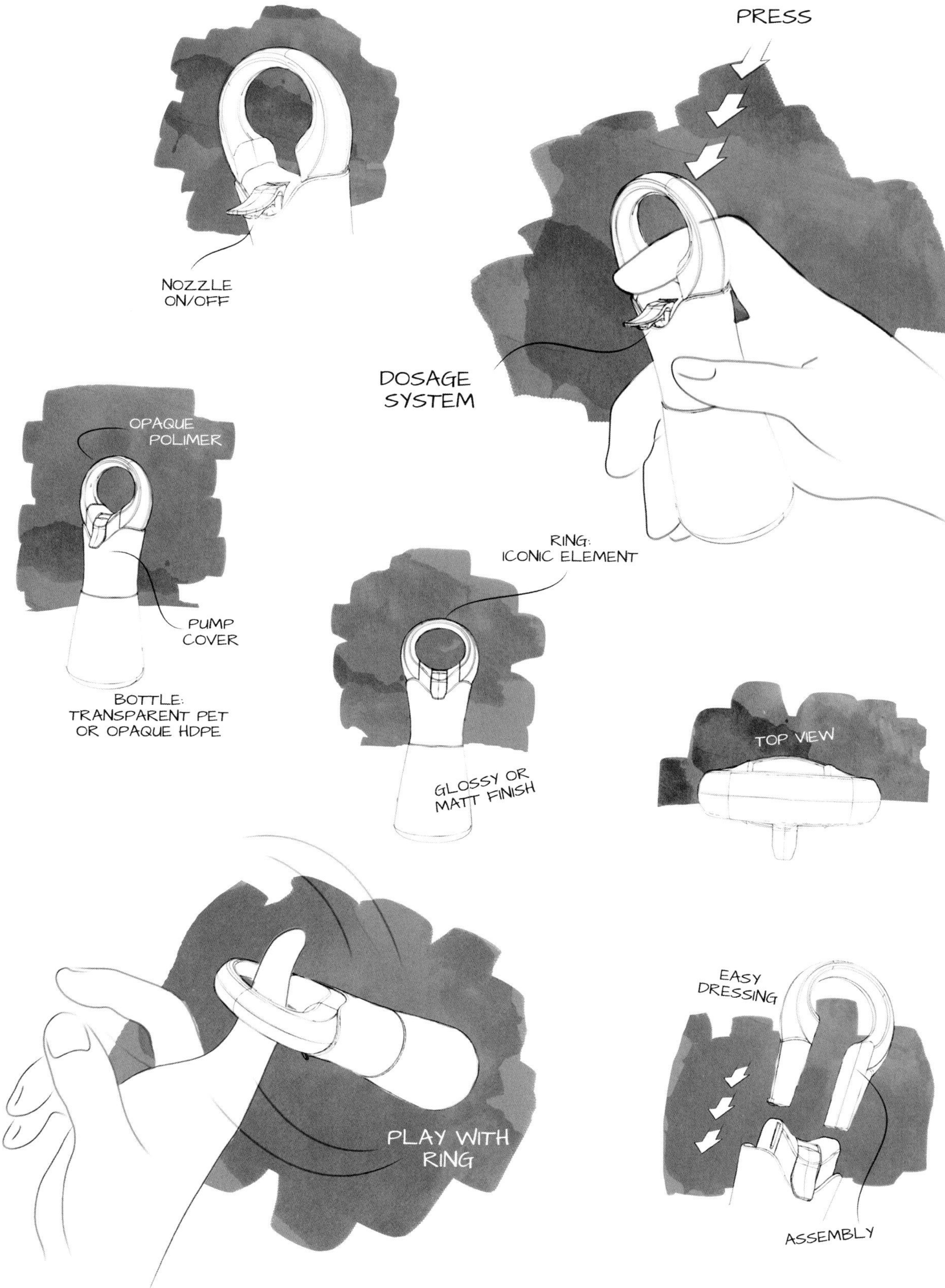
NOZZLE
ON/OFF
PRESS
DOSAGE
SYSTEM
OPAQUE
POLIMER
PUMP
COVER
BOTTLE:
TRANSPARENT PET
OR OPAQUE HDPE
RING:
ICONIC ELEMENT
GLOSSY OR
MATT FINISH
TOP VIEW
PLAY WITH
RING
EASY
DRESSING
ASSEMBLY

Design Agency:
Vozduh Advertising Agency

Creative Director:
Valery Melnik

Designer:
Varya Tcimara

SICO

It is the design of package for a ruler intimate TM SICO gel-greasings.

The decision is based on use of images of characteristic spots in which situations of application of a product are easily guessed.

50 мл
SICO®
гель-смазка
aquagel

50 мл
SICO®
гель-смазка
long action

50 мл
SICO®
гель-смазка
sexyality

50 мл
SICO®
гель-смазка
eroticgel

50 мл
SICO®
гель-смазка
extra safe

FOOD PACKAGE

LAYER HEN

A business souvenir in the form of a limited-edition egg carton, combining the packaging for a dozen eggs with a grocery bag, all decorated in the form of a hen.

The Layer Hen packaging aimed to demonstrate how fresh eggs are and how carefully each of them is produced.

Design Agency:
Geometry Global Moscow

Designer:
Alexandra Istratova

Client:
AB Daily

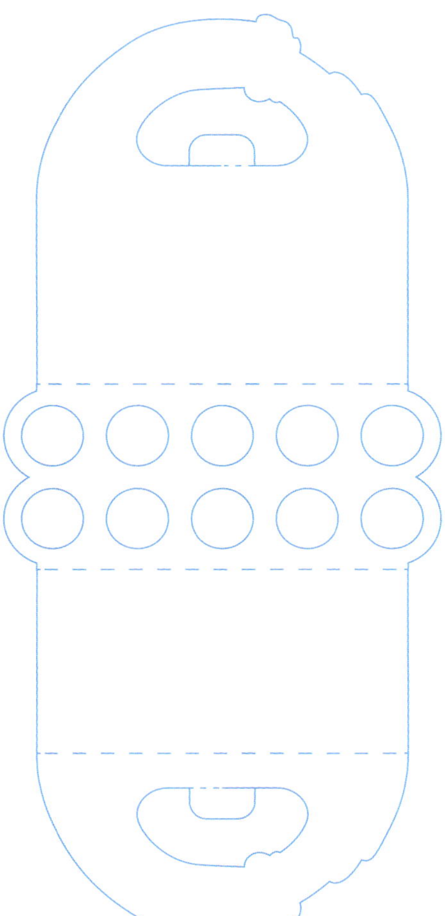

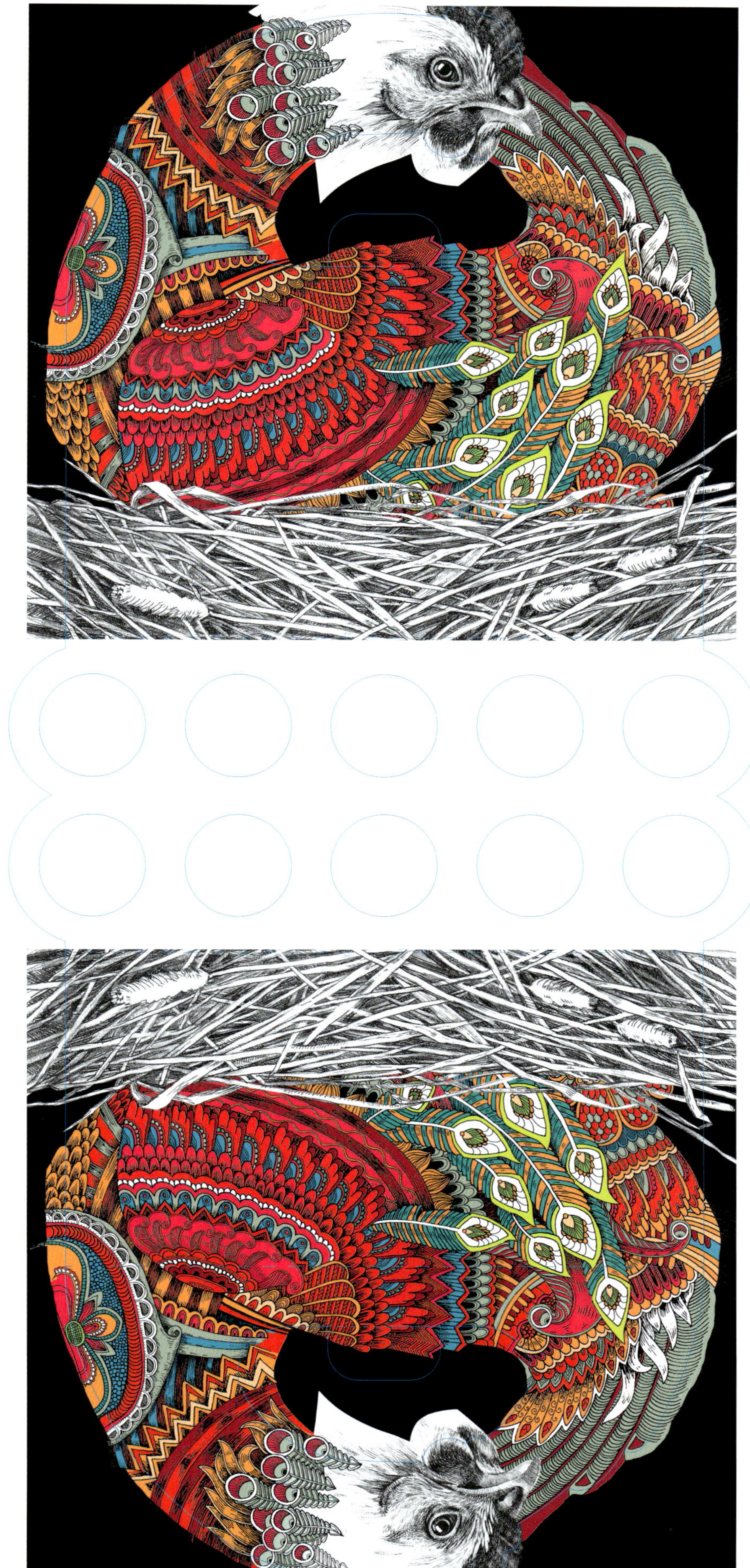

Designer:
Diego Frayle and María Duriana Rodríguez

Photography:
Carlos Arturo Rodríguez
© Master in Packaging Design ELISAVA.
Barcelona, 2014.

AREPA'

The project consisted on creating a packaging system to take away "Arepas" (a traditional and emblematic food of Venezuela and Colombia). Arepa' is for a young audience that has no time and wants to eat well. The proposal consists of a series of folded cardboard packaging without adhesive, manually assembled for easy transportation and consumption of the product. The graphic uses funny characters representing the typical names that have been given to the different types of Arepa's over the years, highlighting their origin and variety of ingredients.

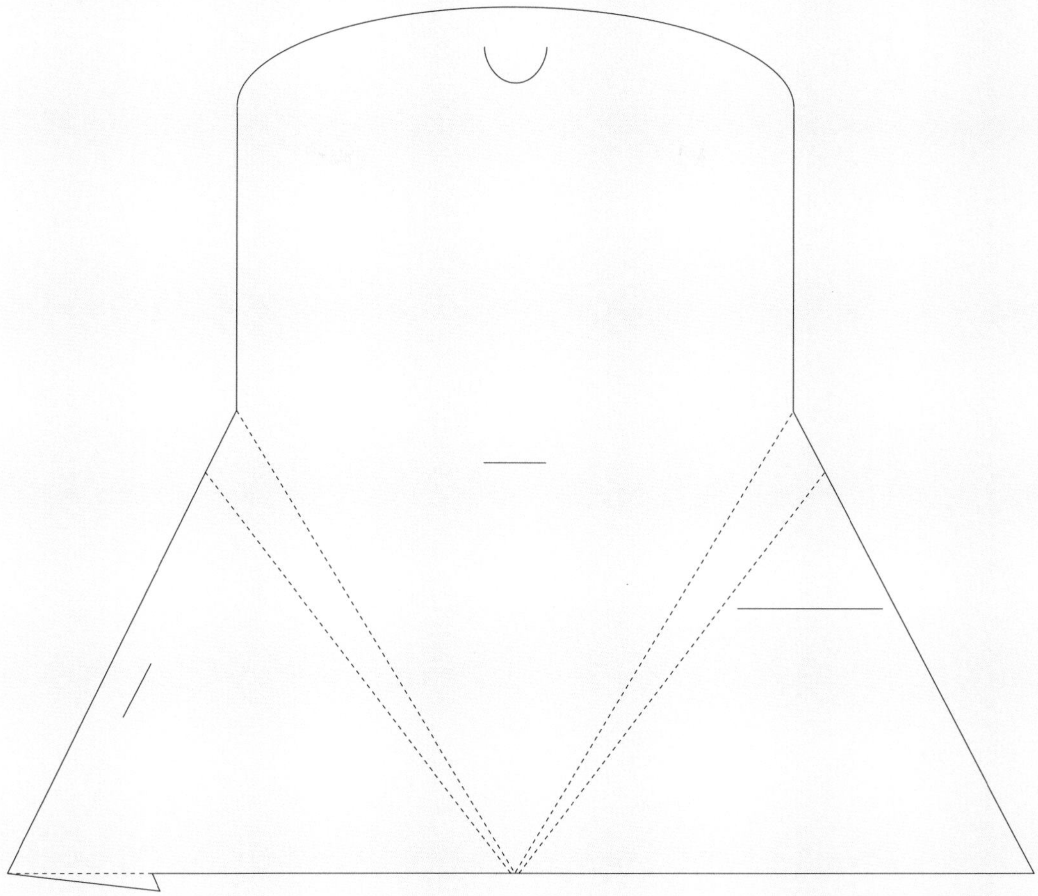

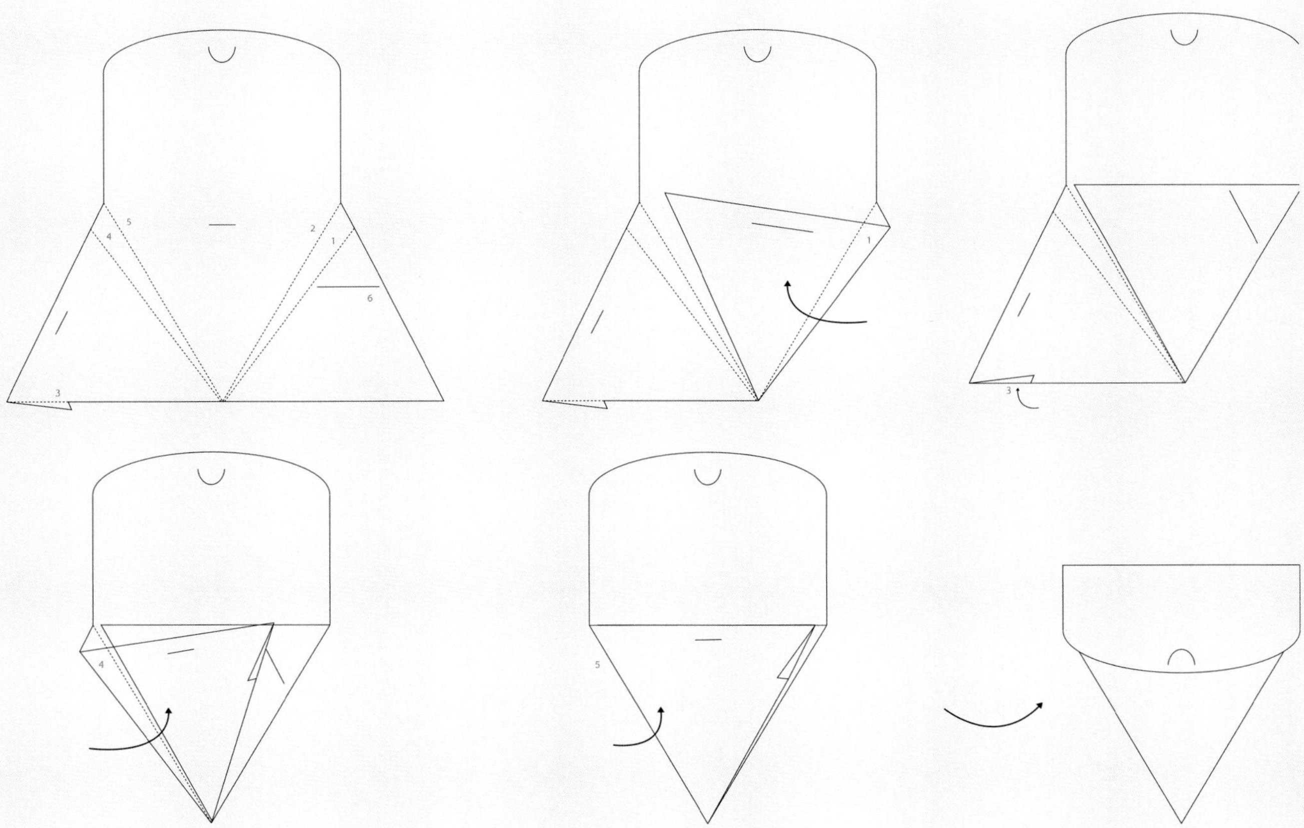

la
RUMBERA

la
PELÚA

reina
PEPIADA

la
DOMINÓ

AURUM

The brand Aurum was created as part of a University assignment and as a pretext to design a visual identity. For this project, I knew that I wanted to work the theme of spices. I therefore created Aurum, a company that produces blends of fine spices.

The brand name was selected due to its meaning, being "gold" in Latin. It has first been chosen for the link that unites spices to wealth and nobility. During the Antiquity, these were so rare and prized that they could easily be compared to gold. The sound of the name Aurum brings to mind the word "aroma", which strengthens the connection with the product even more. In addition, gold is often associated with spices in popular expressions. For example, salt is sometimes called "white gold, and saffron "red gold".

Designer:
Catherine Marion

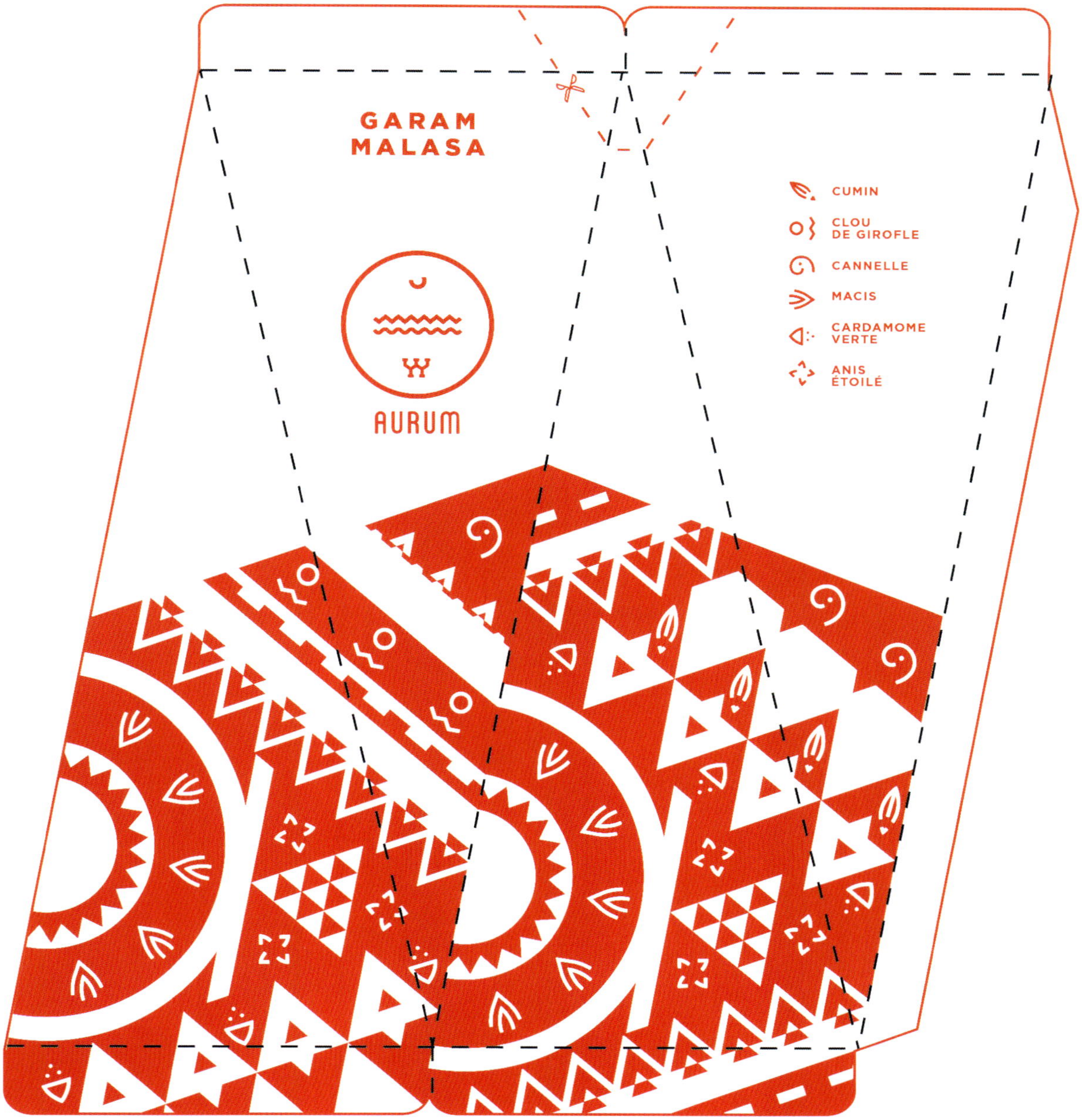
GARAM
MALASA
AURUM
CUMIN
CLOU
DE GIROFLE
CANNELLE
MACIS
CARDAMOME
VERTE
ANIS
ÉTOILÉ

MÉLANGE
QUATRE
POIVRES
AURUM
POIVRE
LONG
POIVRE
CUBÈBE
POIVRE
DE GUINÉ
POIVRE
DE SICHUAN

CINQ
ÉPICES
YUCATÈQUES
AURUM
POIVRE
CUBÈBE
CANNELLE
ORIGAN
MEXICAIN
PIMENTA
DIOICA
CLOU
DE GIROFLE

The logotype itself also has a very symbolic value. A first allusion to gold is made through the use of the circle, since in Alchemy a ring encircling a dot embodies the symbol of gold. Two characters of the Polynesian alphabet were then added within the circle to recall one of the places of origin of spices. The first character takes the shape of the sun, which is often symbolically associated with gold. The second character inspires vegetation, to put forward the natural origin of the product. Finally, in the center of the logotype are an angular and a winding line. The first one reminds the pungent taste of certain aromas but also the piles of spices of Oriental markets. The second one represents sweeter aromas in addition to referring to the Spice Route that was followed by maritime means. The whole creates a logotype whose shape puts emphasis on travel, evoking the stamps found in a passport.

The packaging gives off an exotic feeling by using designs inspired by Maori tattoos, a people of Polynesian origin. It also includes a signs system that was created based on the spices making the blend.

CINQ ÉPICES YUCATÈQUES

ORIGAN MEXICAIN — PIMENTA DIOICA — POIVRE CUBÈBE — CLOU DE GIROFLE — CANNELLE

MÉLANGE QUATRES POIVRES

POIVRE LONG — POIVRE CUBÈBE — POIVRE DE GUINÉ — POIVRE DE SICHUAN

GARAM MALASA

ANIS ÉTOILÉ — CLOU DE GIROFLE — CARDAMOME VERTE — MACIS — CANNELLE — CUMIN

SUNRISE

The project was about developing the packaging for small organic fruits and berries, looking forward to decrease the environmental impact that plastic containers produce.

After analyzing environmental issues and ways of consumption, a cardboard packaging was developed with a structure that does not use glue points; on the other hand, small holes were added to the design in order to create the transparency needed to show fruit when selling. The name was inspired by the sunrise as a synonym for fresh fruits, and the graphics on the labels were based on illustrations that evoke natural and organic origin of fruits to distinguish each product.

Design Agency:
Diego Frayle, María Duriana Rodríguez and Silvia Albertí

Photography:
Carlos Arturo Rodríguez

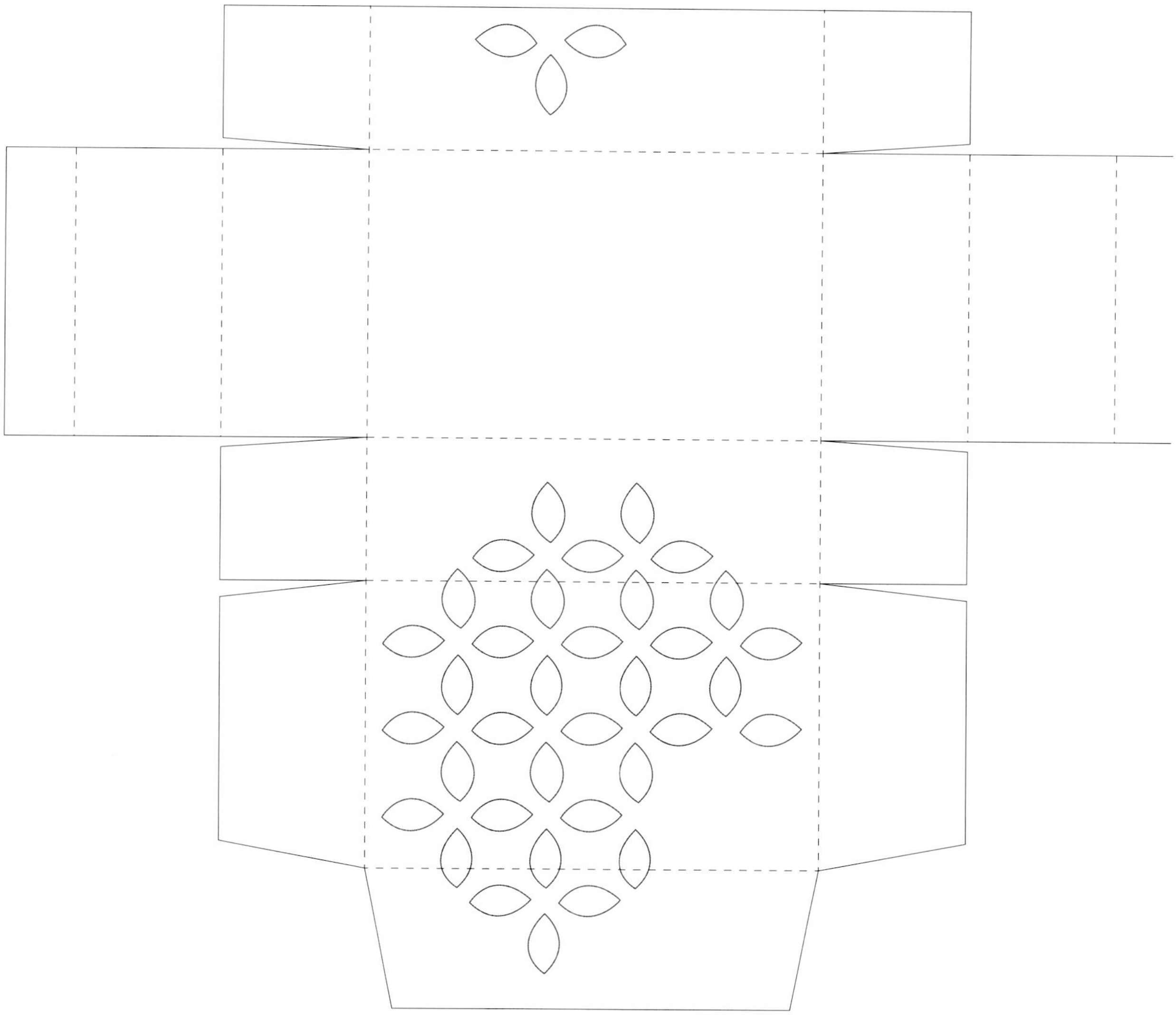

LADOLEA

Our company Melissi&Co produces and distributes, in unique handmade ceramic pots, two superior quality extra virgin olive oils Ladolea (www.ladolea.gr).

The design of our handmade bottle is inspired by the rich history of our Corinthian region in Greece. Dating back to 700 B.C., there was an ancient Corinthian pot, called "Aryballos". This pot was well-known for its use by the athletes, for smearing their bodies with olive oil before the Olympic Games as well as by the aristocracy of the era, for preserving their valuable essential oils and aromas.

Graphic Designer:
Leandros Katsouris

Product Designer:
Spyros Kyzis

Materials:
Stonewear Clay, Cork, Stainless Steel

Contributor:
Melissi&Co

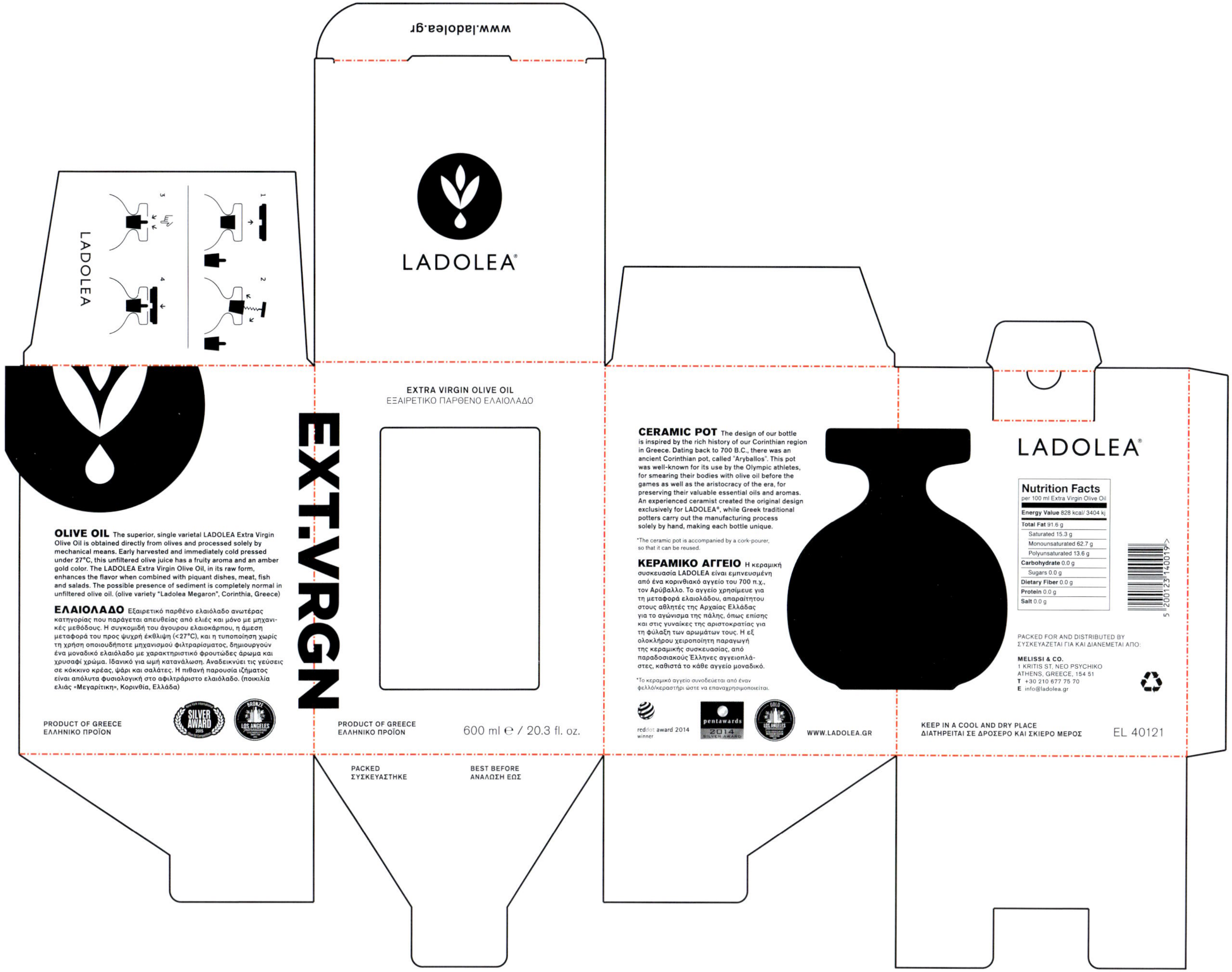
www.ladolea.gr
LADOLEA®
LADOLEA
1
2
3
4
EXTRA VIRGIN OLIVE OIL
ΕΞΑΙΡΕΤΙΚΟ ΠΑΡΘΕΝΟ ΕΛΑΙΟΛΑΔΟ
EXT.VRGN
OLIVE OIL The superior, single varietal LADOLEA Extra Virgin Olive Oil is obtained directly from olives and processed solely by mechanical means. Early harvested and immediately cold pressed under 27°C, this unfiltered olive juice has a fruity aroma and an amber gold color. The LADOLEA Extra Virgin Olive Oil, in its raw form, enhances the flavor when combined with piquant dishes, meat, fish and salads. The possible presence of sediment is completely normal in unfiltered olive oil. (olive variety "Ladolea Megaron", Corinthia, Greece)
ΕΛΑΙΟΛΑΔΟ Εξαιρετικό παρθένο ελαιόλαδο ανωτέρας κατηγορίας που παράγεται απευθείας από ελιές και μόνο με μηχανικές μεθόδους. Η συγκομιδή του άγουρου ελαιοκάρπου, η άμεση μεταφορά του προς ψυχρή έκθλιψη (<27°C), και η τυποποίηση χωρίς τη χρήση οποιουδήποτε μηχανισμού φιλτραρίσματος, δημιουργούν ένα μοναδικό ελαιόλαδο με χαρακτηριστικό φρουτώδες άρωμα και χρυσαφί χρώμα. Ιδανικό για ωμή κατανάλωση. Αναδεικνύει τις γεύσεις σε κόκκινο κρέας, ψάρι και σαλάτες. Η πιθανή παρουσία ιζήματος είναι απόλυτα φυσιολογική στο αφιλτράριστο ελαιόλαδο. (ποικιλία ελιάς «Μεγαρίτικη», Κορινθία, Ελλάδα)
SILVER AWARD 2015
BRONZE LOS ANGELES
PRODUCT OF GREECE
ΕΛΛΗΝΙΚΟ ΠΡΟΪΟΝ
PRODUCT OF GREECE
ΕΛΛΗΝΙΚΟ ΠΡΟΪΟΝ
600 ml ℮ / 20.3 fl. oz.
PACKED
ΣΥΣΚΕΥΑΣΤΗΚΕ
BEST BEFORE
ΑΝΑΛΩΣΗ ΕΩΣ
CERAMIC POT The design of our bottle is inspired by the rich history of our Corinthian region in Greece. Dating back to 700 B.C., there was an ancient Corinthian pot, called "Aryballos". This pot was well-known for its use by the Olympic athletes, for smearing their bodies with olive oil before the games as well as the aristocracy of the era, for preserving their valuable essential oils and aromas. An experienced ceramist created the original design exclusively for LADOLEA®, while Greek traditional potters carry out the manufacturing process solely by hand, making each bottle unique.
*The ceramic pot is accompanied by a cork-pourer, so that it can be reused.
ΚΕΡΑΜΙΚΟ ΑΓΓΕΙΟ Η κεραμική συσκευασία LADOLEA είναι εμπνευσμένη από ένα κορινθιακό αγγείο του 700 π.Χ., τον Αρύβαλλο. Το αγγείο χρησίμευε για τη μεταφορά ελαιολάδου, απαραίτητου στους αθλητές της Αρχαίας Ελλάδας για το αγώνισμα της πάλης, όπως επίσης και στις γυναίκες της αριστοκρατίας για τη φύλαξη των αρωμάτων τους. Η εξ ολοκλήρου χειροποίητη παραγωγή της κεραμικής συσκευασίας, από παραδοσιακούς Έλληνες αγγειοπλάστες, καθιστά το κάθε αγγείο μοναδικό.
*Το κεραμικό αγγείο συνοδεύεται από έναν φελλό/κεραστήρι ώστε να επαναχρησιμοποιείται.
reddot award 2014
winner
pentawards 2014 SILVER AWARD
GOLD LOS ANGELES
WWW.LADOLEA.GR
LADOLEA®
Nutrition Facts
per 100 ml Extra Virgin Olive Oil
Energy Value 828 kcal/ 3404 kj
Total Fat 91.6 g
Saturated 15.3 g
Monounsaturated 62.7 g
Polyunsaturated 13.6 g
Carbohydrate 0.0 g
Sugars 0.0 g
Dietary Fiber 0.0 g
Protein 0.0 g
Salt 0.0 g
5 200123 140019 >
PACKED FOR AND DISTRIBUTED BY
ΣΥΣΚΕΥΑΖΕΤΑΙ ΓΙΑ ΚΑΙ ΔΙΑΝΕΜΕΤΑΙ ΑΠΟ:
MELISSI & CO.
1 KRITIS ST, NEO PSYCHIKO
ATHENS, GREECE, 154 51
T +30 210 677 75 70
E info@ladolea.gr
KEEP IN A COOL AND DRY PLACE
ΔΙΑΤΗΡΕΙΤΑΙ ΣΕ ΔΡΟΣΕΡΟ ΚΑΙ ΣΚΙΕΡΟ ΜΕΡΟΣ
EL 40121

By redesigning this pot, we aspire to bring the archaic to the present and create a modern dispenser that meets the highest market standards, certified by the General Chemical State Laboratory. An experienced ceramist created the original design exclusively for our company, while Greek traditional potters carry out the manufacturing process solely by hand, making each bottle unique.

The bottle is ergonomically designed based on the dimensions of an average user hand and studied through user tests to be re-filled. Due to its special pourer ant lid, no leaks or drops are allowed during serving, offering to the user a functional and clean kitchenware object.

The opaque packaging prevents light penetration, thus maintaining the olive oil in its purest form and preserving the valuable aroma and taste.

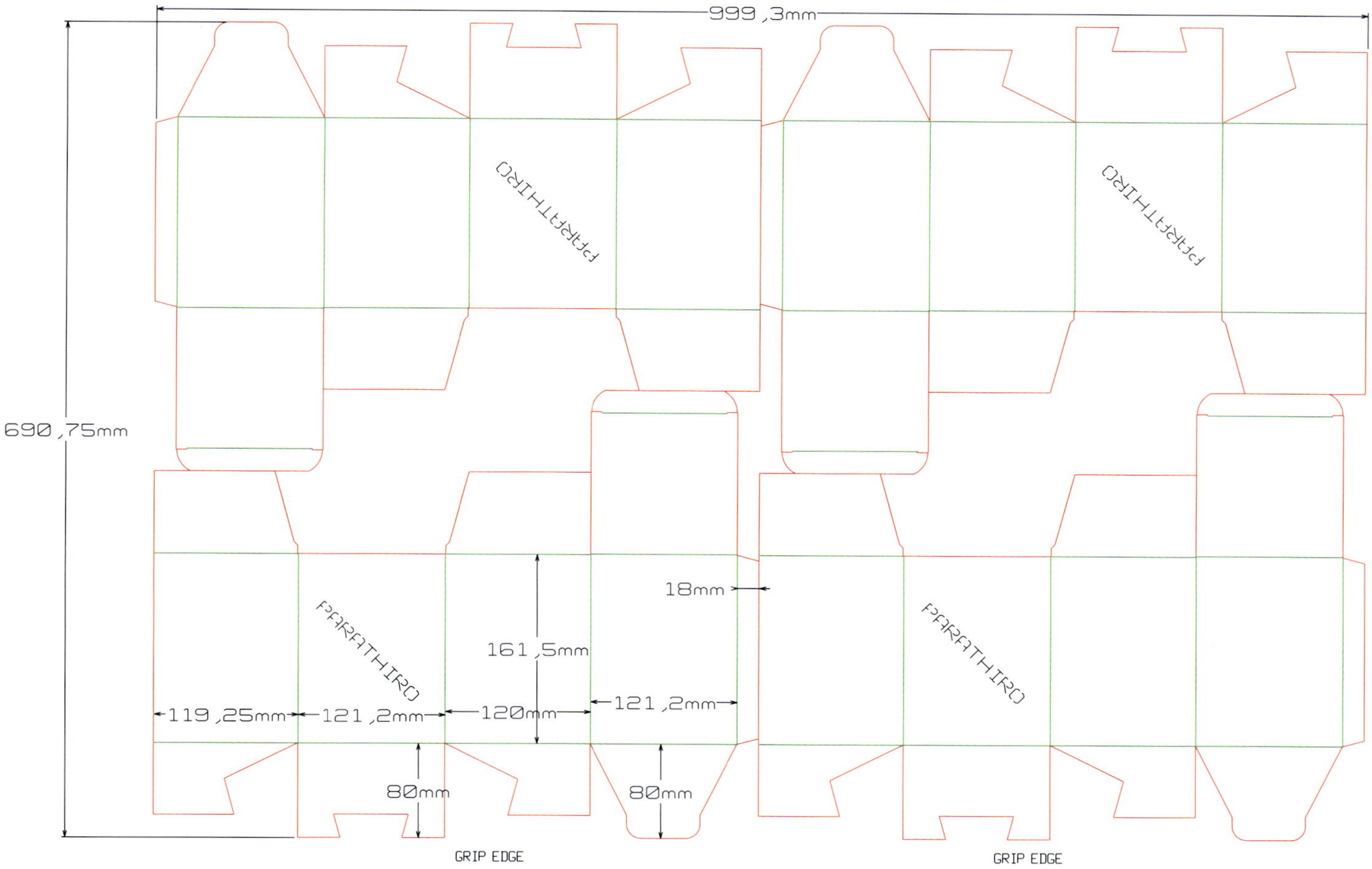
999,3mm
690,75mm
PARATHIRO
PARATHIRO
PARATHIRO
PARATHIRO
18mm
161,5mm
119,25mm
121,2mm
120mm
121,2mm
80mm
80mm
GRIP EDGE
GRIP EDGE

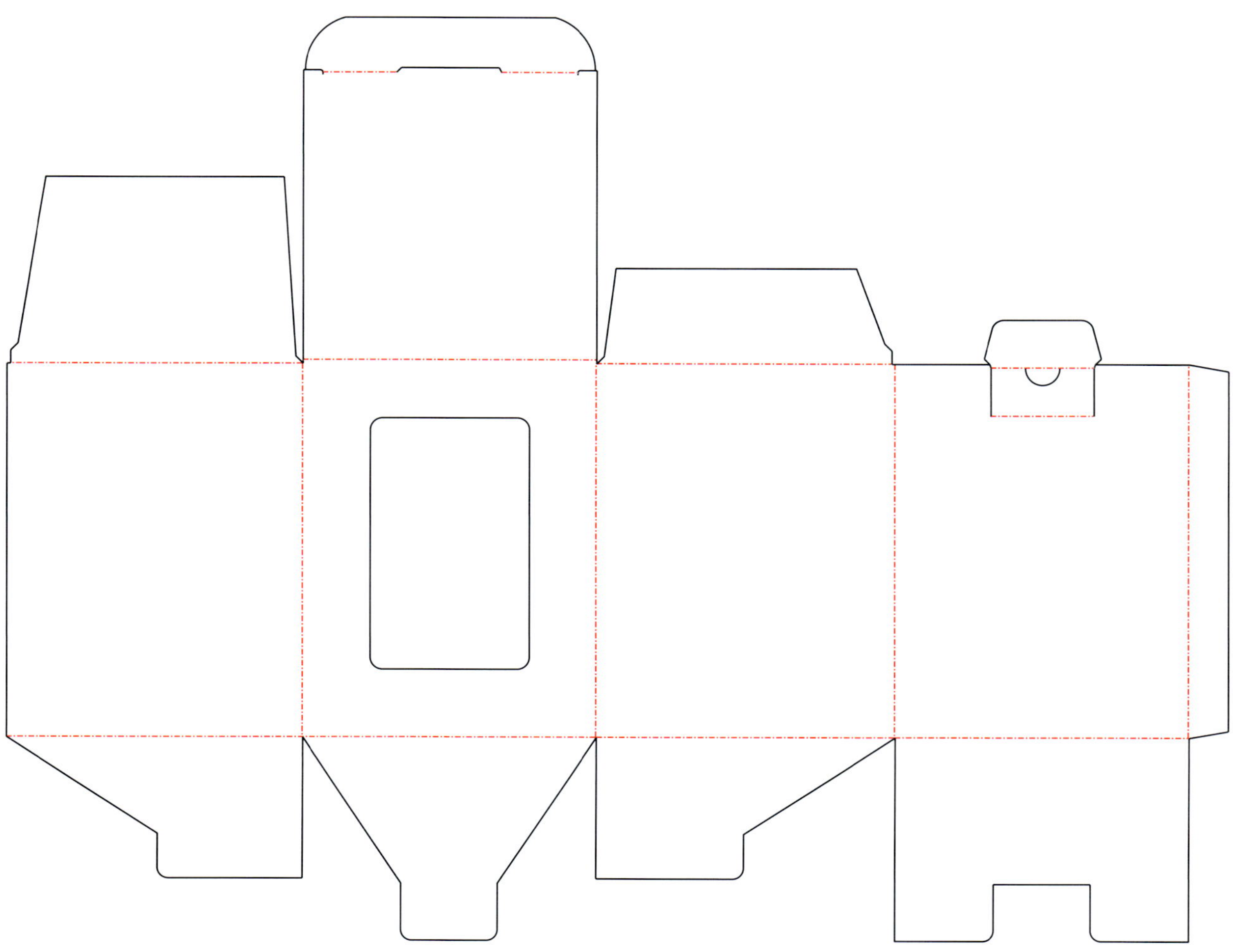

TOKYO SASHISUSESO

"Sa-Shi-Su-Se-So" is a wordplay used in Japan to make it easier to remember the five basic seasonings in Japanese cuisine and the correct order for using them when cooking ("sa" for satou or sugar, "shi" for shio or salt, and so on for vinegar, soy sauce and miso). "Tokyo Sashisuseso" is a brand for the five traditional seasonings in the Tokyo area. The symbol of the brand is composed of traditional forms, taking inspiration from the cherry flower that is emblematic of the city of Tokyo, as well as from emblems that were historically used by merchant families. The five seasonings come together as petals that form a cherry flower, expressing the fact that they are seasonings of Tokyo. The lateral sides of the package provide the impression of woodgrain and are inspired by the forms of traditional Japanese boxes.

The package has a built-in handle string that allows it to be carried. The labels are printed not with ink but with black silver leaves that endow the design with a stately elegance.

Art Director:
Yukihiko Aizawa

Director:
China Nagata

Designer:
Yukihiko Aizawa

Client:
Food Relation Co., Ltd

TOKYO SASHISUSESO
TRADITIONAL SEASONING OF TOKYO

CONDIMENT SET

"Packaging design for selected condiments for the company Spice Pinch." This is Bachelor Thesis done at the Academy of Art and Design in Wroclaw. The main goal of my project was to create a packaging, which could be also used as a condiment set. It would be easy to use and it would have unique shape, characterized by the simplicity of execution and clear graphics. While I was designing shape of the package I thought about scattered spices on the table, that form hillock. That thought led me to design a package that resembles a hillock - a package in form of a triangle.

Designer:
Adrianna Broniszewska

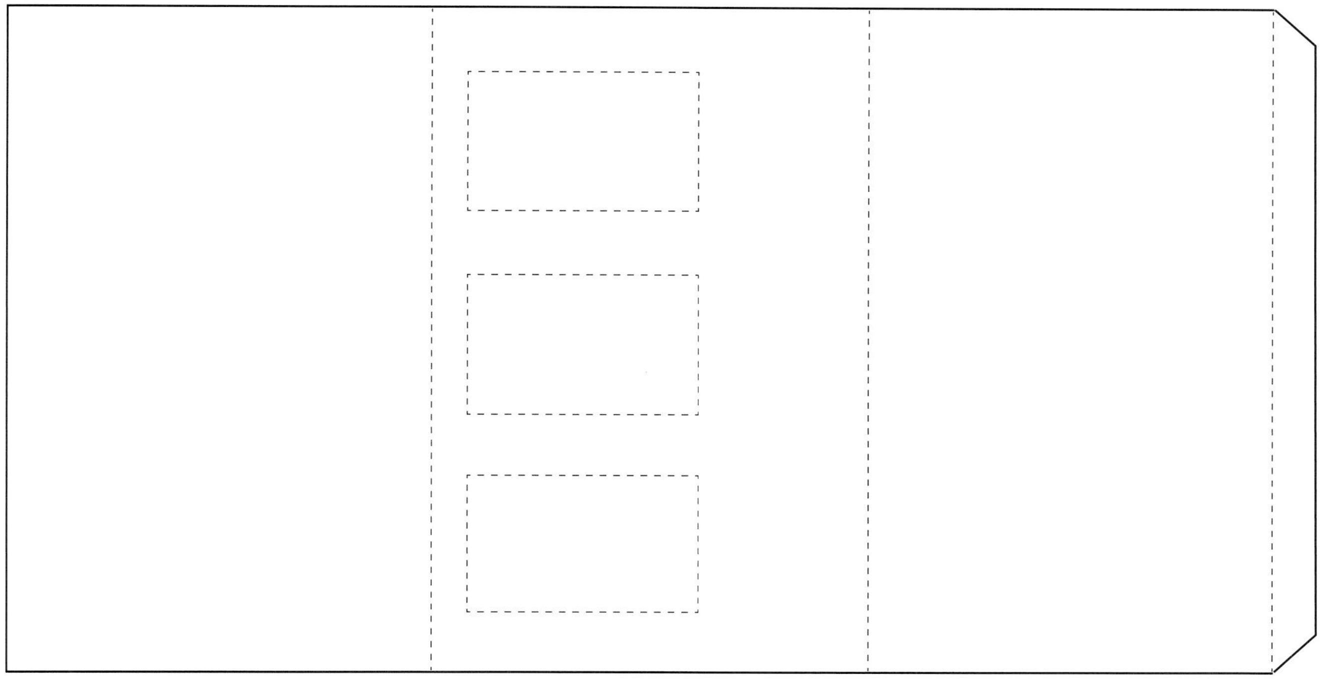

Kminek
PRZYPRWA
ZIOŁOWA
SZCZYPTA PRZYPRAW
Papryka
OSTRA MIELONA
PRZYPRWA
WARZYWNA
Pieprz
CZARNY MIELONY
PRZYPRWA
KORZENNA
Pieprz
CZARNY ZIARNISTY
PRZYPRWA
KORZENNA
Bazylia
PRZYPRWA
ZIOŁOWA

Bazylia
PRZYPRWA
ZIOŁOWA
PPPBKGGG

Kminek
PRZYPRWA
ZIOŁOWA
SZCZYPTA PRZYPRAW

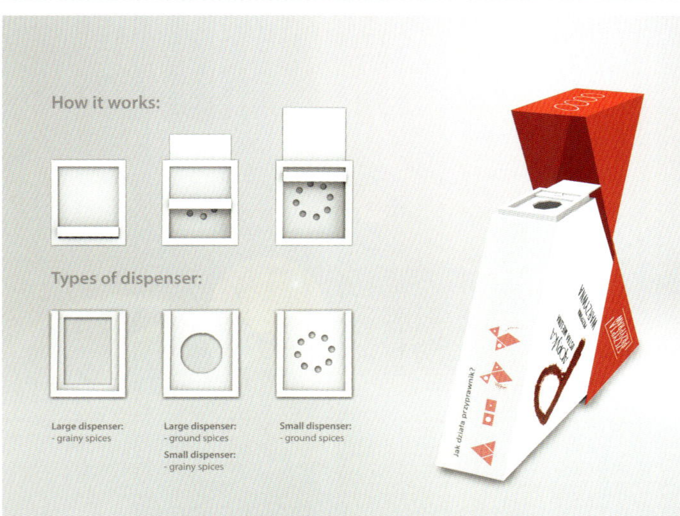
How it works:
Types of dispenser:
Large dispenser:
- grainy spices
Large dispenser:
- ground spices
Small dispenser:
- grainy spices
Small dispenser:
- ground spices

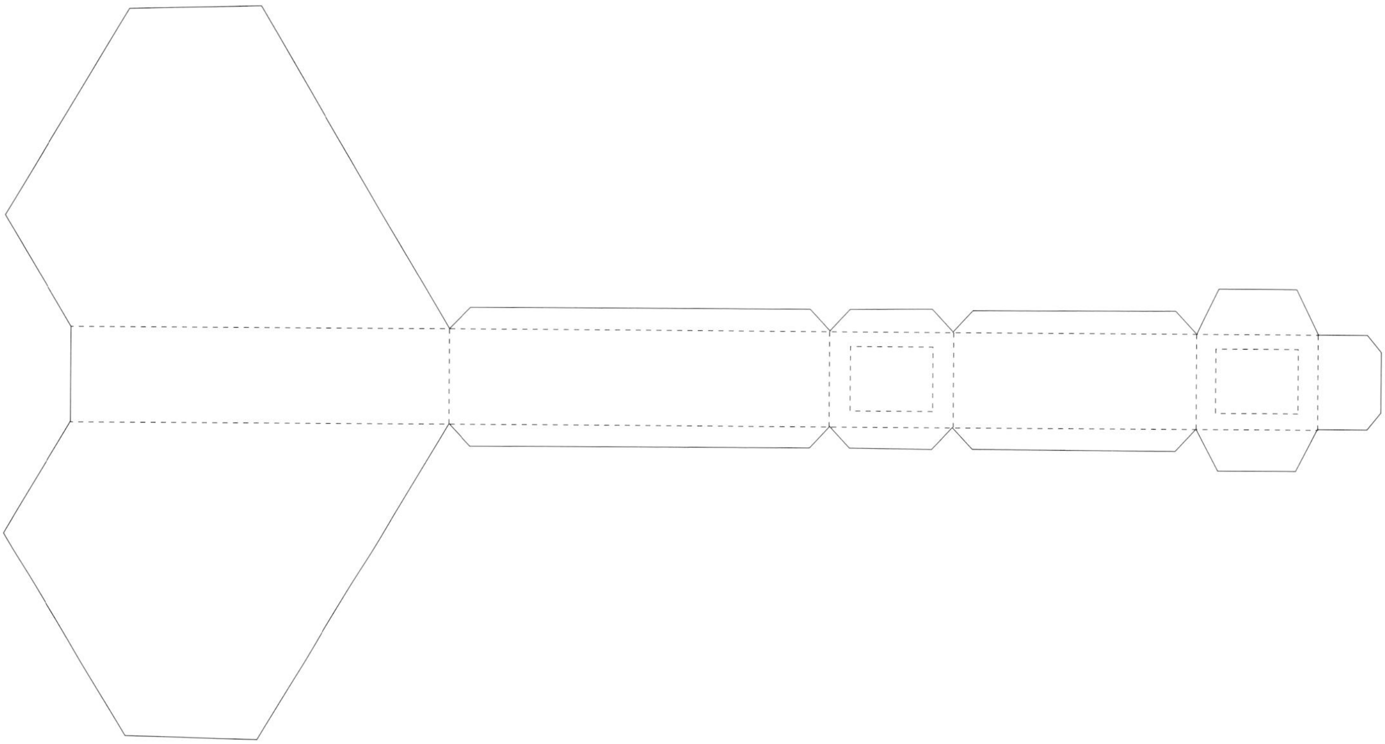

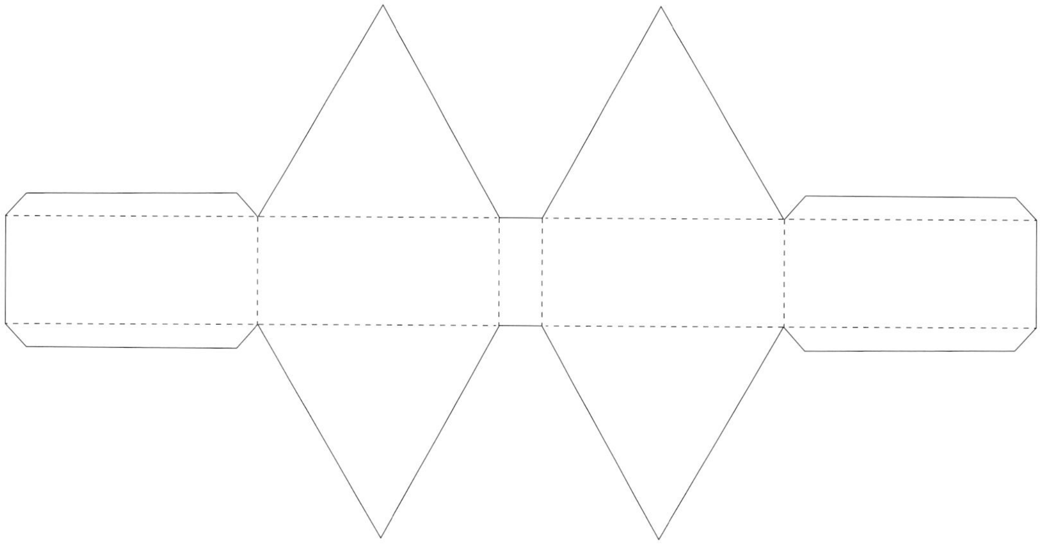

Designer:
Christian Pannicke

Photography:
Christian Pannicke

RHONE

The Rhône is a more than 800 km long river that flows in France. The river offers perfect conditions for growing wine, herbs, various fruits and olive trees. The ragged mountain landscape meets bright light coming from the south.

These facts form the basis for Rhone, a fictional manufacturer of the finest olive oil. Rhone combines classic elegance with modern clarity. The logo is a combination of word and figurative mark. A classic typographic approach was combined with a modern, reduced illustration. A clear distinction from competitors' products is generated by dispense with typical images of olives, olive trees or landscapes. Instead, the scenic contrast is used in dark mountains and glittering sunlight and translated into a monochrome design. The packaging is reduced to the essentials and attracts attention especially because of this contrast.

Rhone embodies enjoyment and delight, as well as an extraordinary quality of life. The hard contrast, as well as the landscape, is broken by a pleasant, bright and colorful imagery.

RHONE EXTRA VIRGIN
COLD EXTRACTION OLIVE OIL

Rhone

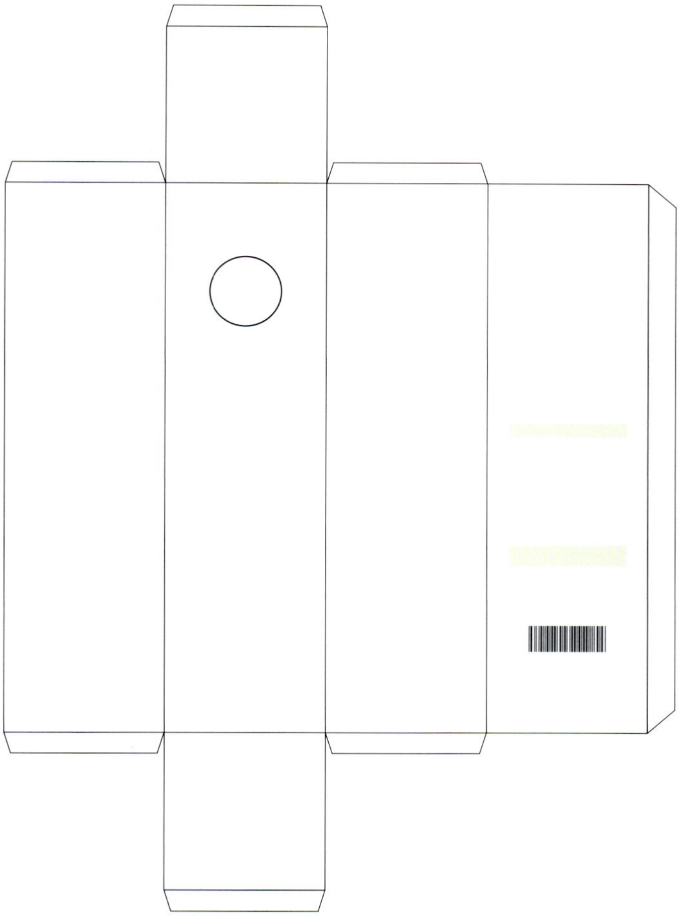

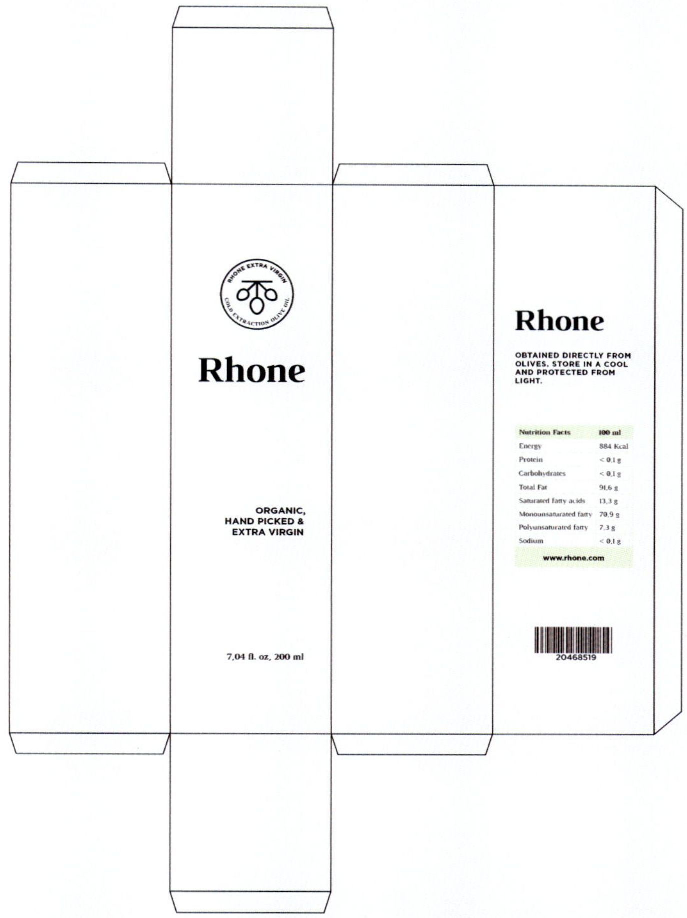

RHONE EXTRA VIRGIN
COLD EXTRACTION OLIVE OIL
Rhone
ORGANIC,
HAND PICKED &
EXTRA VIRGIN
7,04 fl. oz, 200 ml
Rhone
OBTAINED DIRECTLY FROM OLIVES. STORE IN A COOL AND PROTECTED FROM LIGHT.
Nutrition Facts 100 ml
Energy 884 Kcal
Protein < 0,1 g
Carbohydrates < 0,1 g
Total Fat 91,6 g
Saturated fatty acids 13,3 g
Monounsaturated fatty 70,9 g
Polyunsaturated fatty 7,3 g
Sodium < 0,1 g
www.rhone.com
20468519

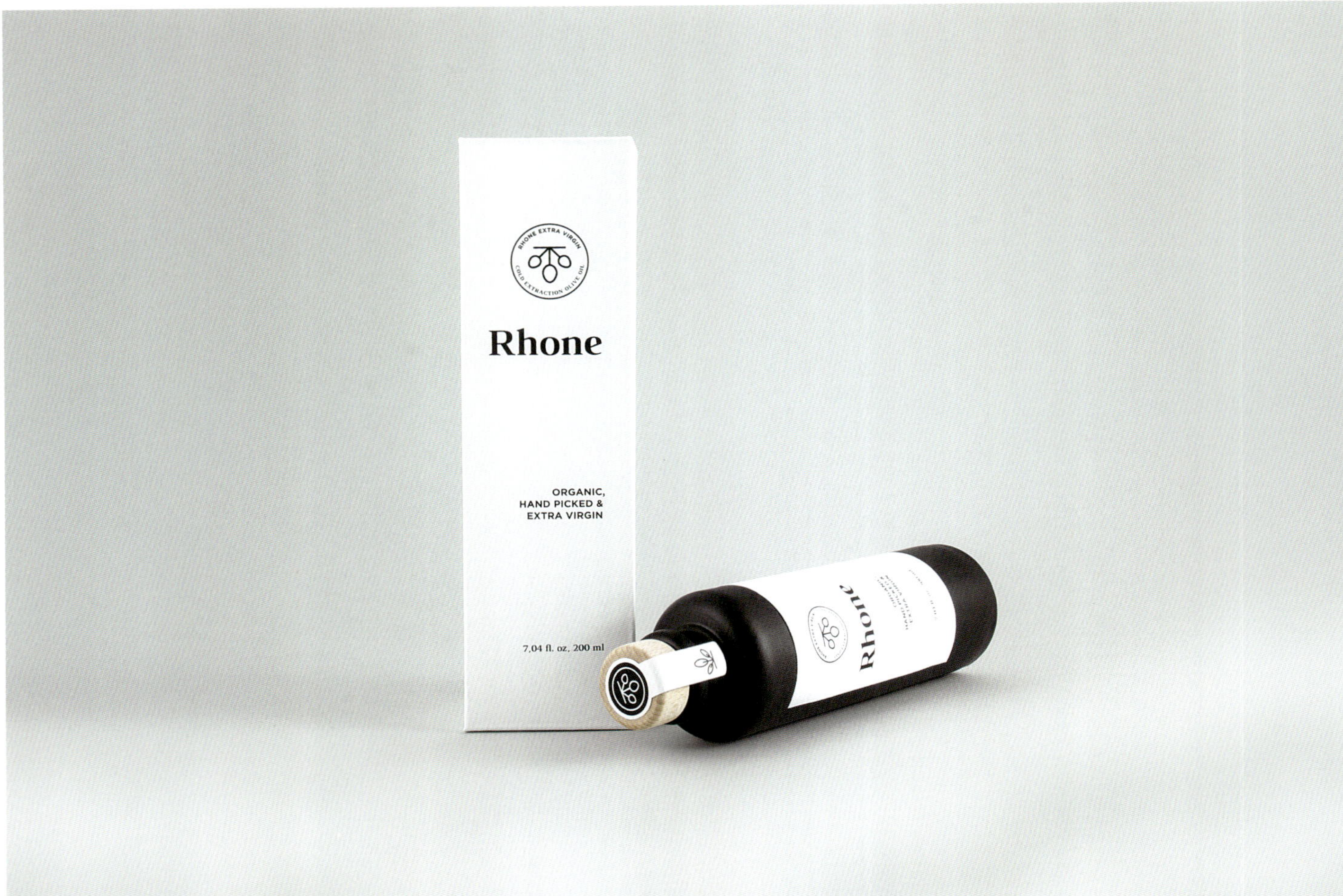
RHONE EXTRA VIRGIN
COLD EXTRACTION OLIVE OIL
Rhone
ORGANIC,
HAND PICKED &
EXTRA VIRGIN
7,04 fl. oz. 200 ml
Rhone

RHONE EXTRA VIRGIN
COLD EXTRACTION OLIVE OIL
Rhone
ORGANIC,
HAND PICKED &
EXTRA VIRGIN
7,04 fl. oz. 200 ml
Rhone
OBTAINED DIRECTLY FROM OLIVES. STORE IN A COOL AND PROTECTED FROM LIGHT.
Nutrition Facts 100 ml
Energy 884 Kcal
Protein < 0,1 g
Carbohydrates < 0,1 g
Total Fat 91,6 g
Saturated fatty acids 13,3 g
Monounsaturated fatty 70,9 g
Polyunsaturated fatty 7,3 g
Sodium < 0,1 g
www.rhone.com

RHONE EXTRA VIRGIN
COLD EXTRACTION OLIVE OIL

Rhone

**ORGANIC,
HAND PICKED &
EXTRA VIRGIN**

7,04 fl. oz, 200 ml

Rhone

OBTAINED DIRECTLY FROM OLIVES. STORE IN A COOL AND PROTECTED FROM LIGHT.

Nutrition Facts	100 ml
Energy	884 Kcal
Protein	< 0,1 g
Carbohydrates	< 0,1 g
Total Fat	91,6 g
Saturated fatty acids	13,3 g
Monounsaturated fatty	70,9 g
Polyunsaturated fatty	7,3 g
Sodium	< 0,1 g

www.rhone.com

MOOLOKO

There's a line of milk and dairy product called "MOOLOKO" ("MOO" + "Moloko" which means milk" in Russian). The best way to show a cow is to put fingers at your head and say "MOO", so I've decided to use these folds in the package to represent horns and put "MOO" in a shape of cow's nose. Its production was in December 2014.

The new brand of dairy products is to be introduced in Tomsk, a city in Siberia, Russia, home of Tomsk State University and students' capital. The brand is marketed as mid-premium, and targeted to young and hip audience.

The name came from the client, they've decided to come up with 'МУЛОКО' ('MOOLOKO'), which stands for Russian word for milk 'МОЛОКО' ('MOLOKO'), but the first syllable was changed to 'МУ' ('MOO').

I've decided to use both specifics of the package which was predefined by the client, and also to work with the naming. So, the whole pack represents the cow's head, flaps turn into ears/horns when opened, and the black nose contains the big 'MOO' in it. At the very end, when you open your pack to enjoy your milk — the cow metaphorically moves its ears and says 'MOO!'

I have chosen monochrome to keep it simple and focused, and also to provide space for possible line extensions — for sour milk, cottage, ryazhenka etc, to be used in different colours. So thus we help consumer to differentiate between different products within one line!

MOO!

Art Director and Designer:
Klemenkov Andrey aka Cocha & Bepeck

3D Visualization:
Massimo Brugnera

Client:
OOO "Derevenskoe molochko", Tomsk, Russia

PURE-PAK®
ELOPAK
НАЖАТЬ
ELOPAK
НАЖАТЬ
ОТКРЫВАТЬ ЗДЕСЬ

ALMA DE CUBA

In the 1940s, Cuba was the world's biggest exporter of coffee, shipping 20,000 tonnes of beans every year. But then came the Revolution and with it a sure and steady decline in production. By 2012, Cuban coffee, once the world's favourite, was on its knees, and it looked like nothing could save it.

Well, that was before Phillip Oppenheim – a former MP and card-carrying Cubaphile – got together with some old pals to restore Cuban coffee to its rightful place. He asked us to help with the branding, but we were so impressed, we ended up investing in the business and becoming partners, too.

We came up with the name Alma de Cuba (soul of Cuba) and channelled the spirit of Cuban folk art (via our resident Latin American illustration genius) to create an evocative design scheme which, hopefully, has just as much soul as the coffee itself.

Design Agency:
Big Fish

Creative Director:
Perry Haydn Taylor

Designer:
Ariel Cortese

¡Tómelo!
con amor de Cuba
CAFÉ
¡Fuerte!
ALMA DE Cuba

GRAGNANO - PASTA ASCIUTTA ITALIANA

Gragnano is a Brazilian brand of pasta that intends to bring the Italian tradition to the table of Brazilians. With handmade features, Gragnano seeks to be a premium brand that offers a quality product to its customers, inspired by the Italian cuisine. Patterns of pasta and split syllables printed onto kraft tube identify the type of pasta. Texture of kraft paper and the badge logo ensure a more artisanal aspect, emphasizing it was handling with care and love.

Design Agency:
Okta Design

Designer:
Camila Chisini

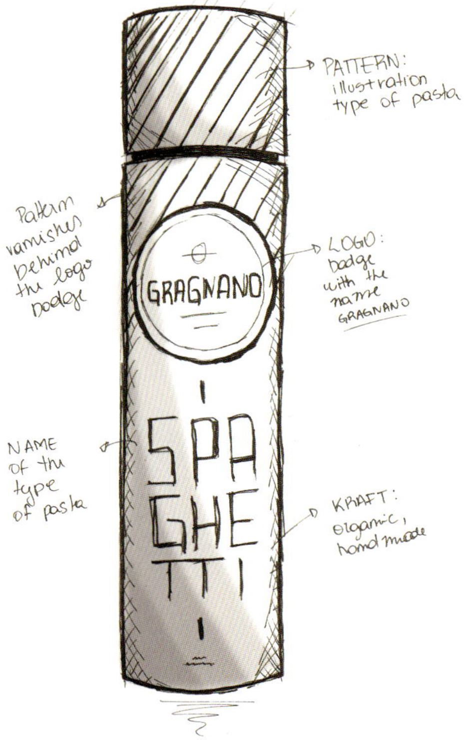

COVER

COVER SUPPORT

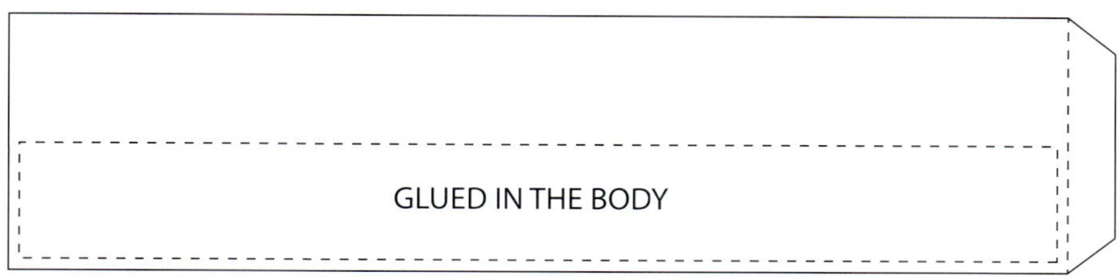

BODY

TOP

BOTTOM

Design Agency:
Reverse Innovation

Creative Director:
Alice Tacconi

Graphic Designer:
Fiona Martin

GIOIELLO DI SAPORE

Excellence, preciousness and Italian good taste: these three concepts were the inspiration for the New Year's gift sent by Reverse Innovation to their clients and partners. A small square glass jar holds a delicious gourmet orange preserve, packaged in special box with an unusual structure. The graphic design focuses on the baroque heritage of Sicily, an Italian region that is famous for its sunshine, cuisine and rich culture. The "jewellery-box" pack elegantly combines an intense black colour with satin silver, to communicate the precious nature of the gift. Once the top is removed, the base opens like a flower and the glass jar is revealed surrounded by a circular baroque motif in UV varnish. The cool metallic silver contrasts with the warm tones of the fruit. As a final touch, a message of good wishes can be seen only when the jar is lifted up.

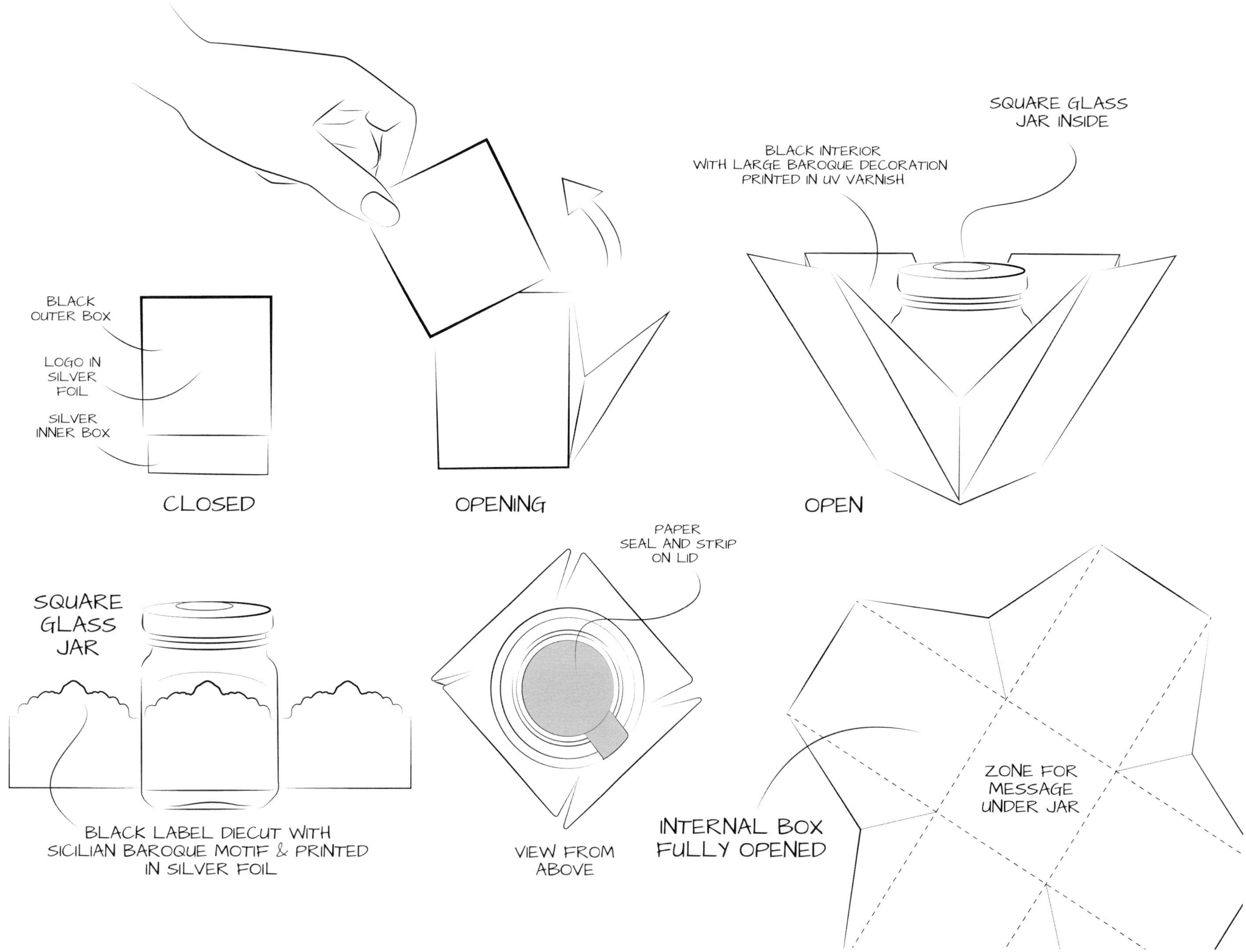
BLACK OUTER BOX
LOGO IN SILVER FOIL
SILVER INNER BOX
CLOSED
OPENING
BLACK INTERIOR WITH LARGE BAROQUE DECORATION PRINTED IN UV VARNISH
SQUARE GLASS JAR INSIDE
OPEN
SQUARE GLASS JAR
BLACK LABEL DIECUT WITH SICILIAN BAROQUE MOTIF & PRINTED IN SILVER FOIL
PAPER SEAL AND STRIP ON LID
VIEW FROM ABOVE
INTERNAL BOX FULLY OPENED
ZONE FOR MESSAGE UNDER JAR

HONEY PACKAGING

Honeycombs are seemingly simple, yet intricate structures that are created by honeybees. The bees don't really get enough credit for the beautiful structures they construct. My concept behind this packaging project focuses on the structure of honeycomb, and highlights the beauty of both the honey and the hexagon itself. For the logotype, I replaced the traditional "O" with a hexagon that affects the surrounding characters. This hexagonal shape is then repeated through the use of die cuts around the jar, as well as the shape of the jar itself. Because of this, the jars of honey can be stacked in a honeycomb pattern. The windows that are cut around the jar are intended to showcase the elegant, golden honey and comb that are held within.

Designer:
Derek Dubler

Photography:
Derek Dubler

BUTTER!

This packaging design for butter is a combination of the conventional paper packaging and the butter can. That's how it combines the advantages of the two types of packaging: It is stable, clean, reclosable and can be used perfectly not just at the table, but also for cooking.

All the packages have in common that they are user-optimized and easy to handle. Four packages are made with folding techniques; one package has a fixed shape.

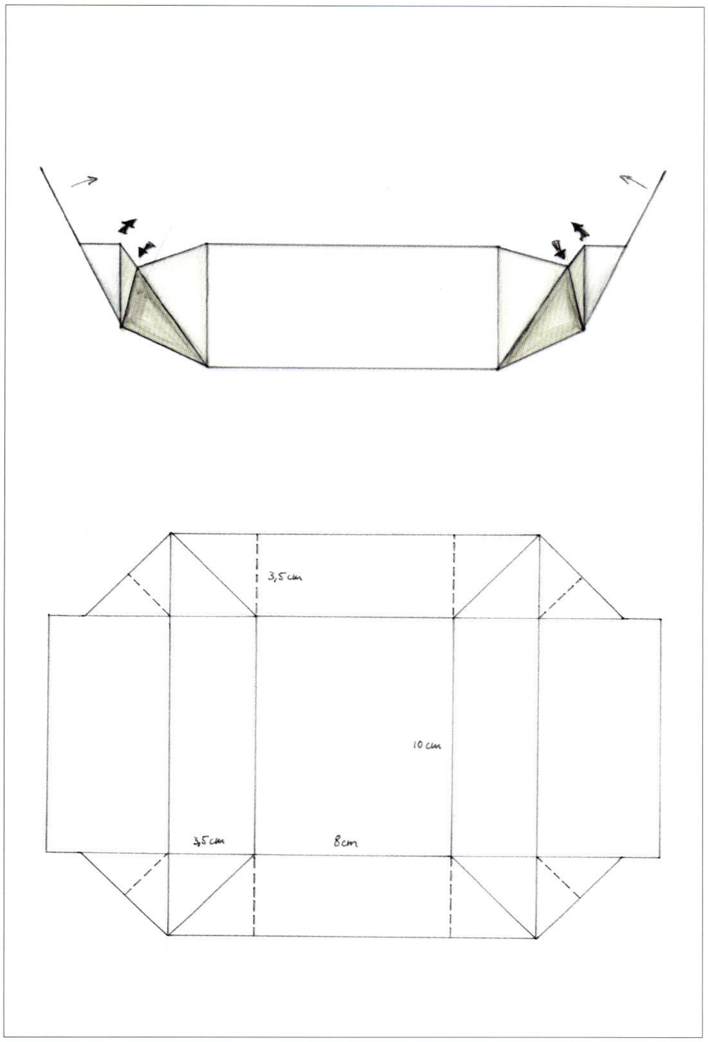

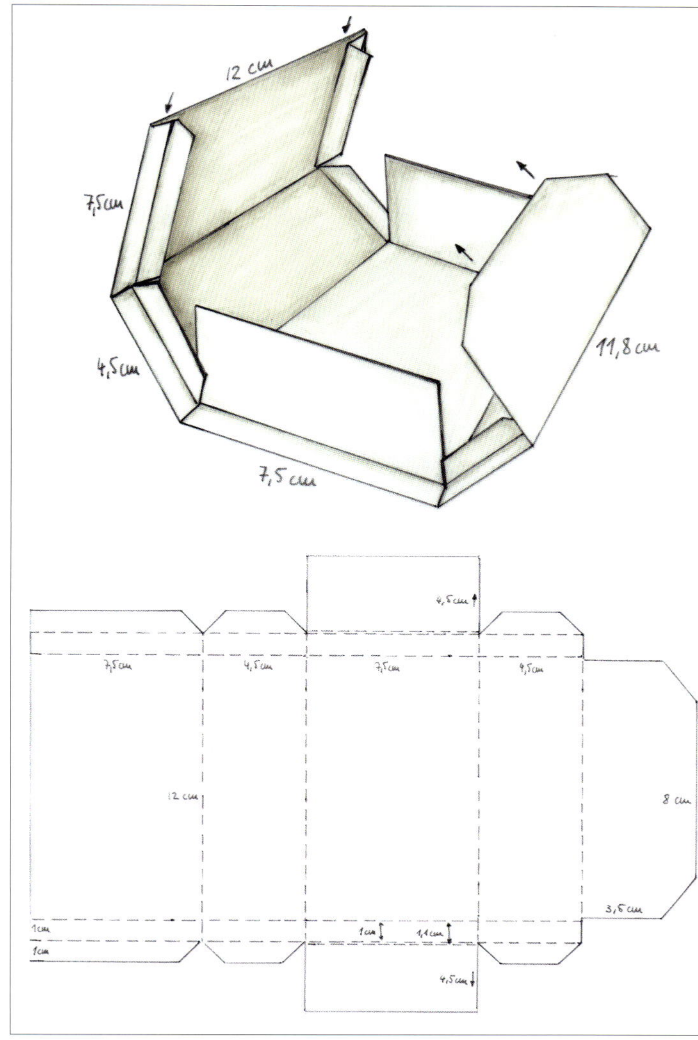

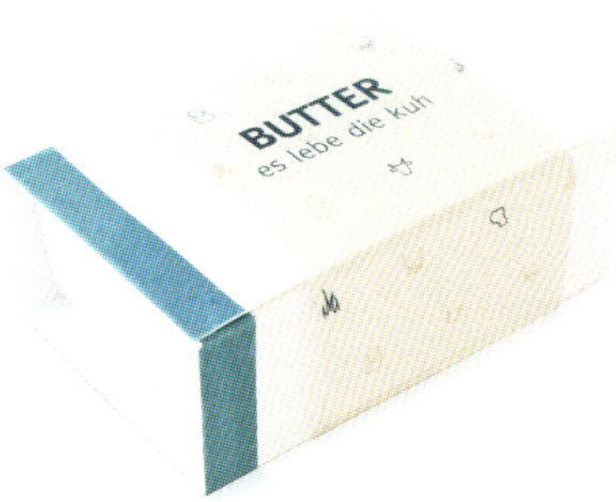

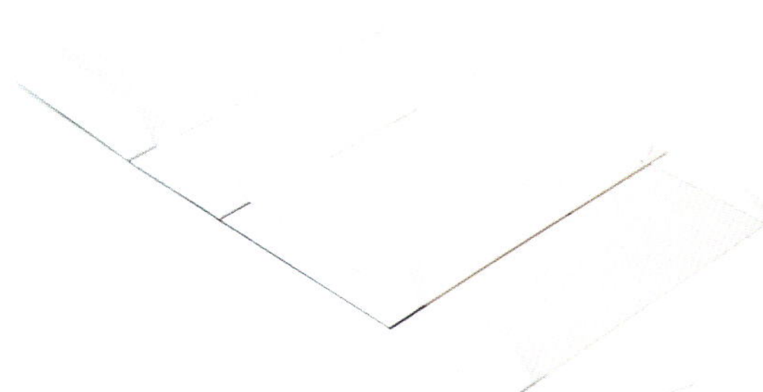

Designer:
Mara Holterdorf & Alessia Sistori

BUTTER
es lebe die kuh

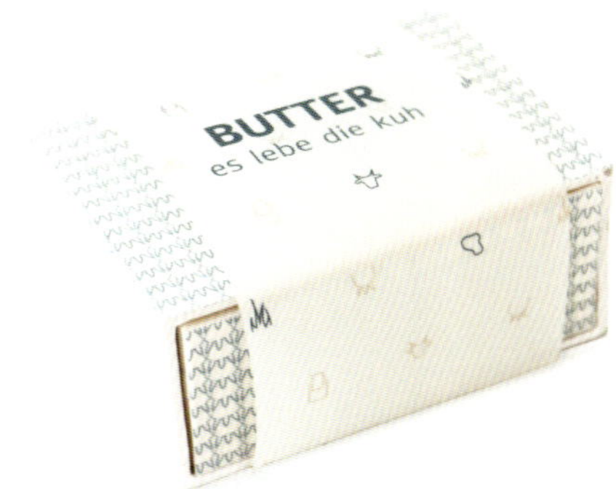
BUTTER
es lebe die kuh

BUTTER
es lebe die kuh

BUTTER

BUTTER
es lebe die kuh

BUTTER

BUTTER
es lebe die kuh

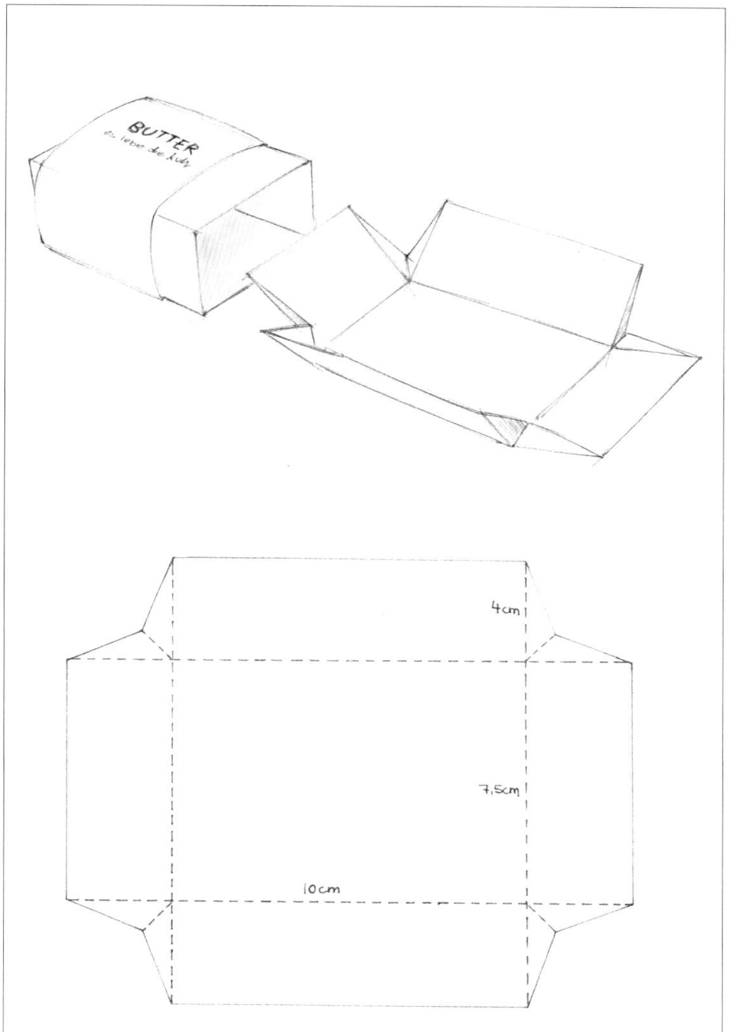
BUTTER
4cm
7,5cm
10cm

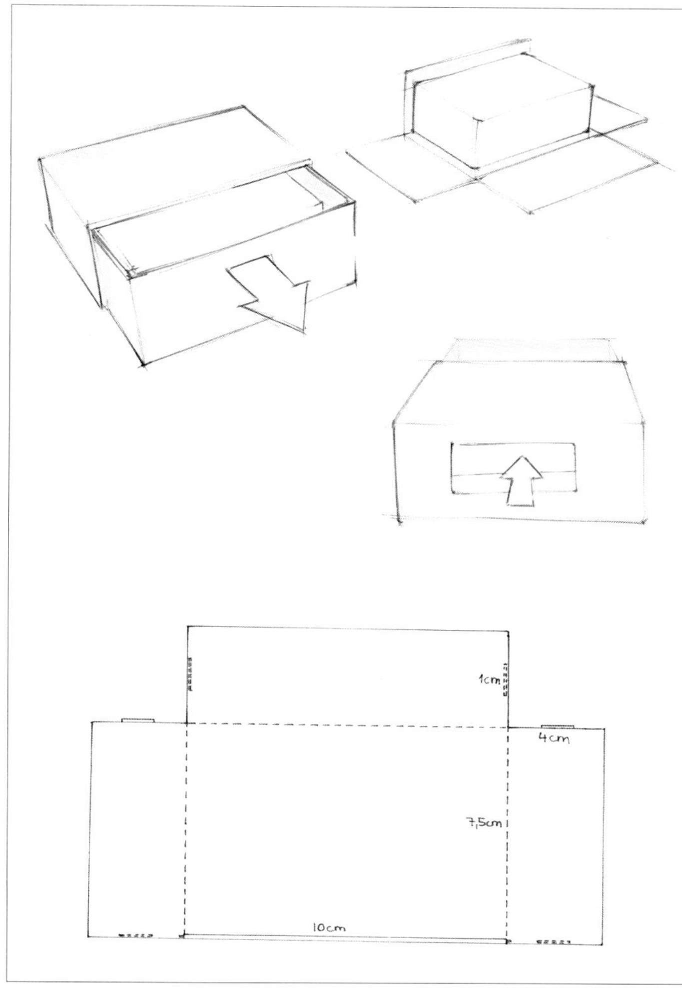
1cm
4cm
7,5cm
10cm

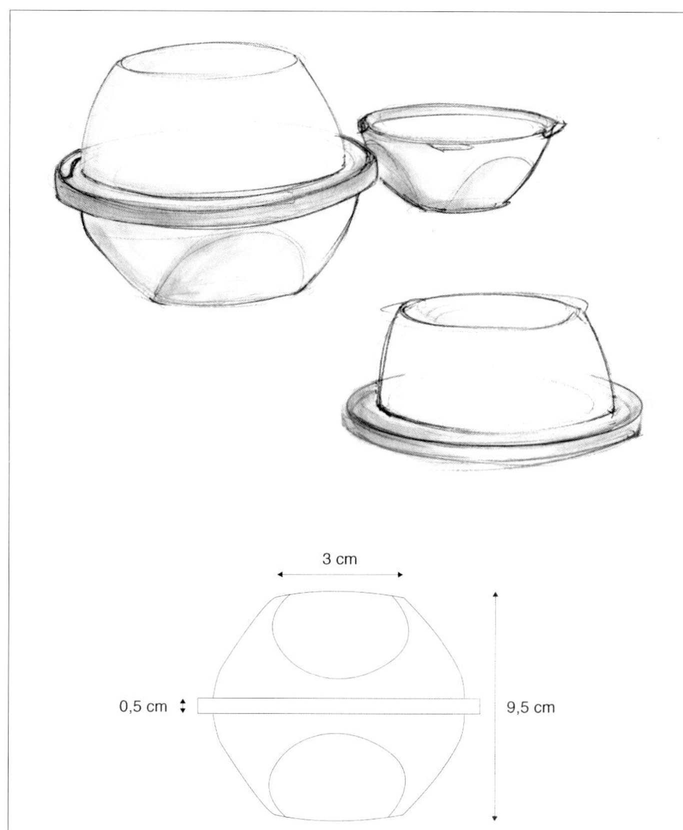
3 cm
0,5 cm
9,5 cm

Designer:
Gustav Karlsson

Photography:
Gustav Karlsson

KINO TAKE AWAY

Kino is a combined vegetarian restaurant, bar and cinema in Gothenburg, Sweden.

These are three takeaway boxes, with integrated handles. One for the main course, one for soups and stews and one for salad/bread. Inspired by vintage camera bags, the packages have an inside print/pattern for that pleasant surprise. The material is a sustainable sourced paperboard with a starch lining, 100% compostable and microwave-friendly.

The top is both easy to close and open. There is a perforation around the package, so it quickly transforms to a plate.

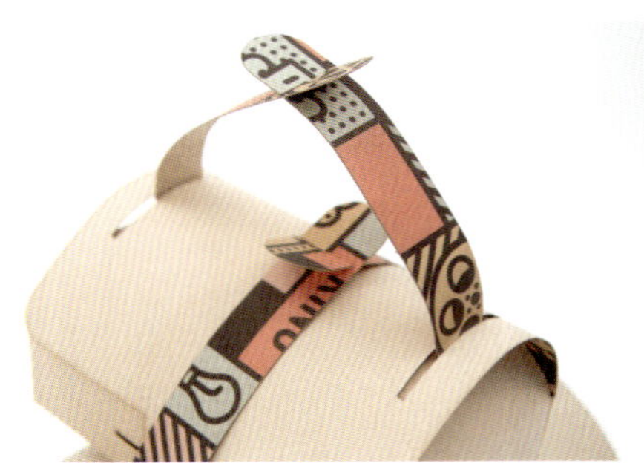

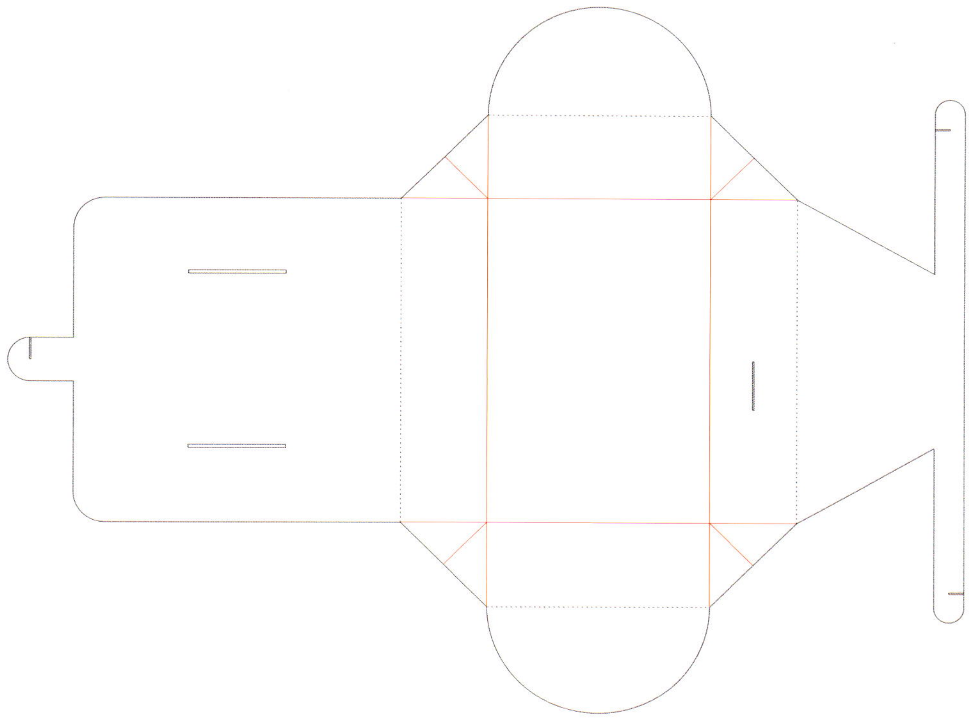

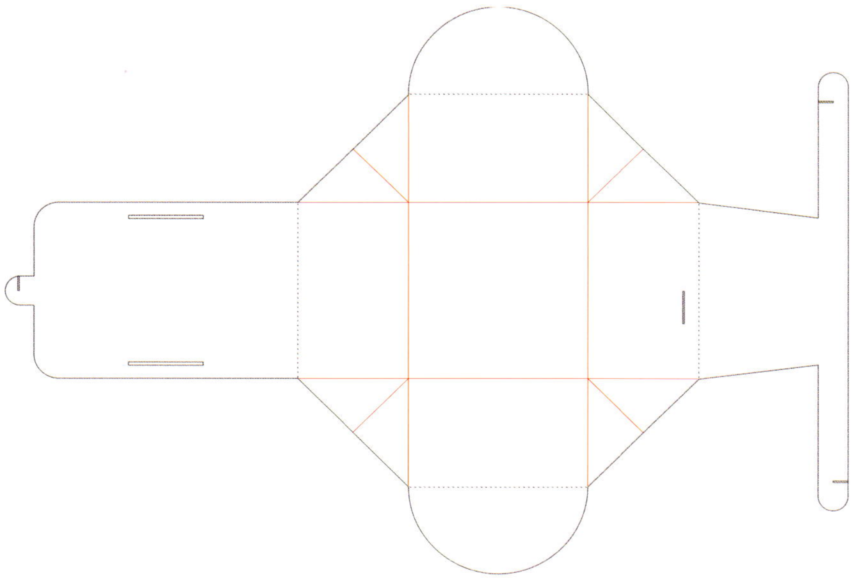

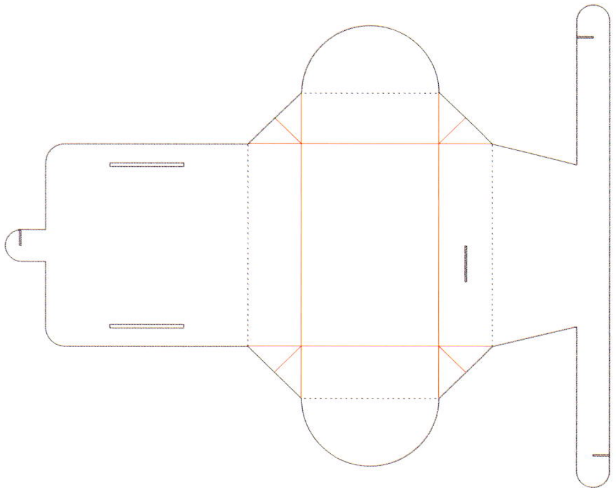

Design Agency:
MATE Studio

Designer:
Alice Macarova

Client:
MATE studio

Photography:
Alice Macarova

MARINE SWEETS

The name "Marine sweets" is very well-known in Russia. This product has a lot of designs solutions.

So my studio decided to create something new and unusual for it. In this project I tried to connect detailed illustrations with minimalistic design.

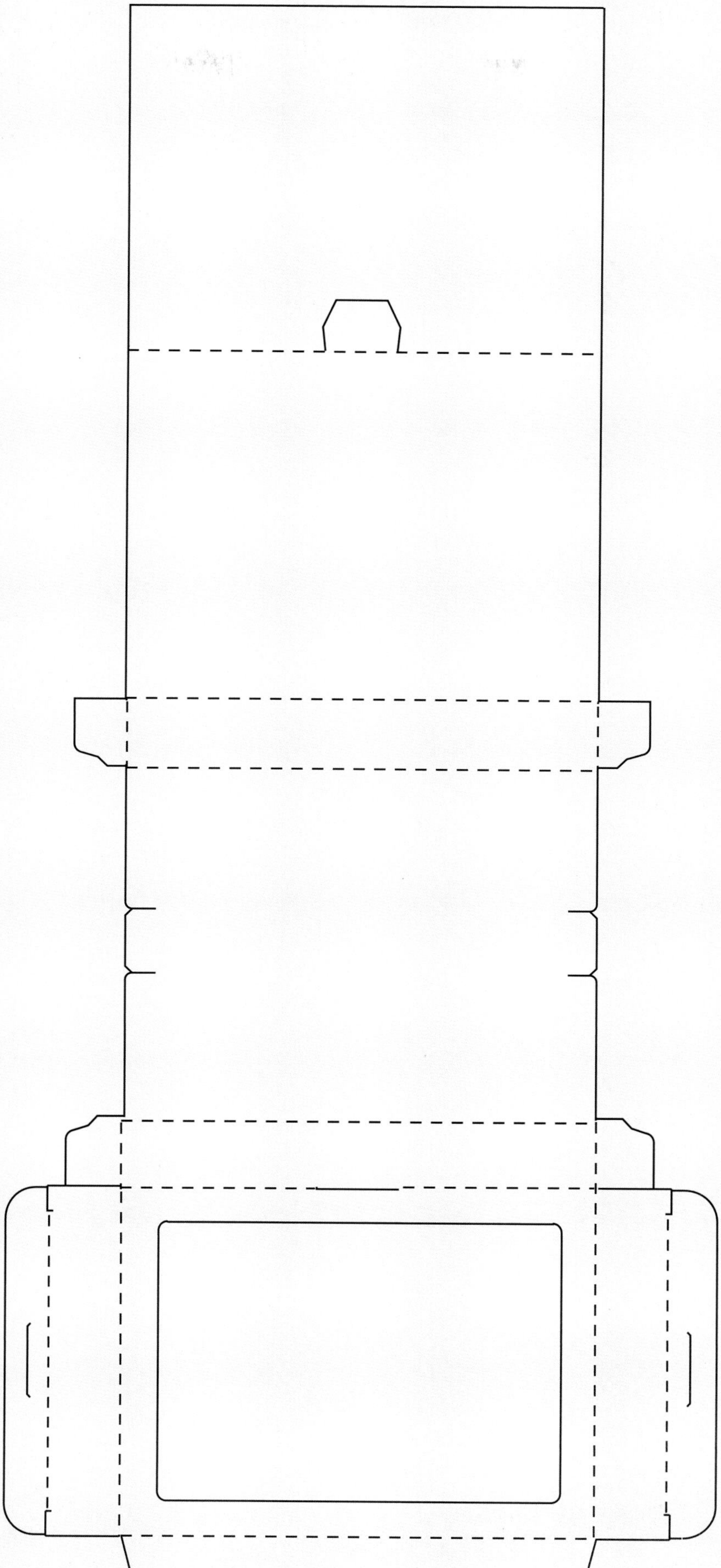

ANGUS 6

Angus 6 is a premium beef brand that sells only Australian beef. We worked with our client from setting the concept of the brand, package design, catalog design, and for the website design as well. The concept behind this brand was "The Great Inheritance," to emphasize the great nature of Australia. We wanted to give exquisite high quality brand look by using strong black and white color palette, and spare use of beefy red. The traditional serif typeface (both English and Korean) adds authentic look to the brand as well.

Design Agency:
Eggplant Factory

Creative Director:
Jeyoun, Lee

Designer:
Bobae, Kim / Youngji, Jung

Food Stylist:
Euna, Kim

Client:
Angus 6

Photography:
Seunghui, Lee

ANGUS 6
NATURE GUARANTEES THE QUALITY
MASTERPIECE
WELL MADE
ANGUS 6
NATURE GUARANTEES THE QUALITY

NADA

Nada isn't just a brand for healthy food, but a brand that communicates through fashion, beauty, well being and body confidence. These core values reflect across the brands packaging and visual identity, through the use of minimalism contrasting with an array of bright and geometrical patterns. Minimalism has been applied to the lid of the packaging, which represent the products value of innocence, health and well being. The patterns that have been applied to the container symbolises fashion, beauty, the fusion of flavours and various pasta / noodle shapes.

Designer:
Thomas Squire & Eve Warren

Client:
Disclosed

Photography:
Eve Warren & Thomas Squire

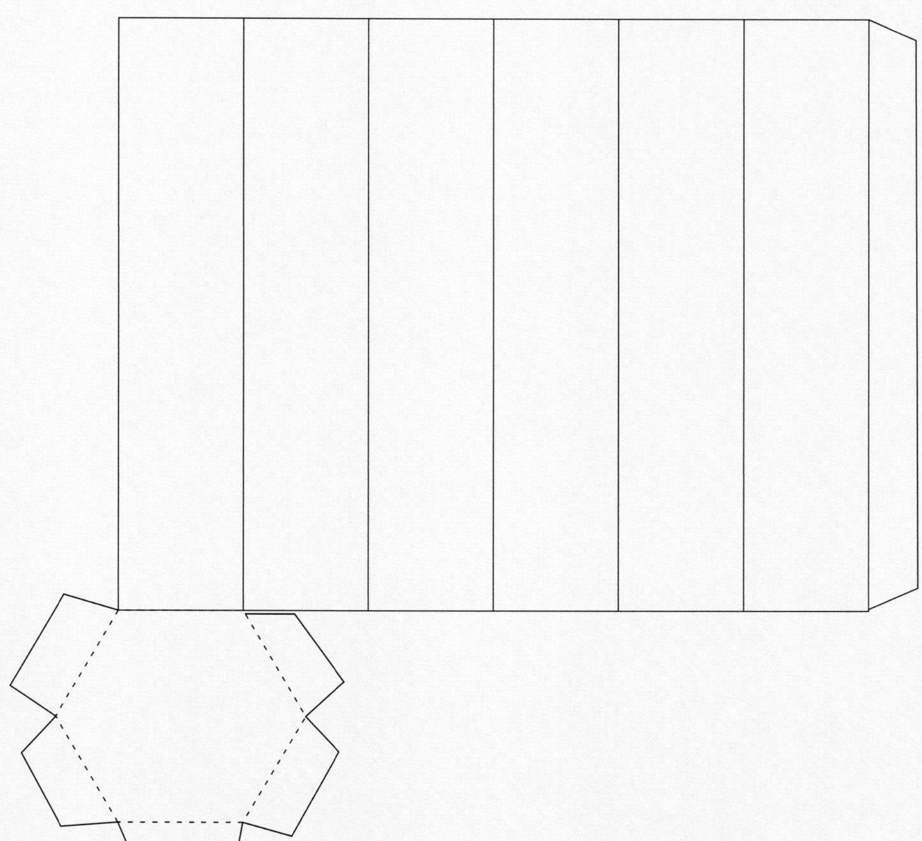

DELICIOUS

500g ℮
NADA
Sweet Potato & Buckwheat Noodles

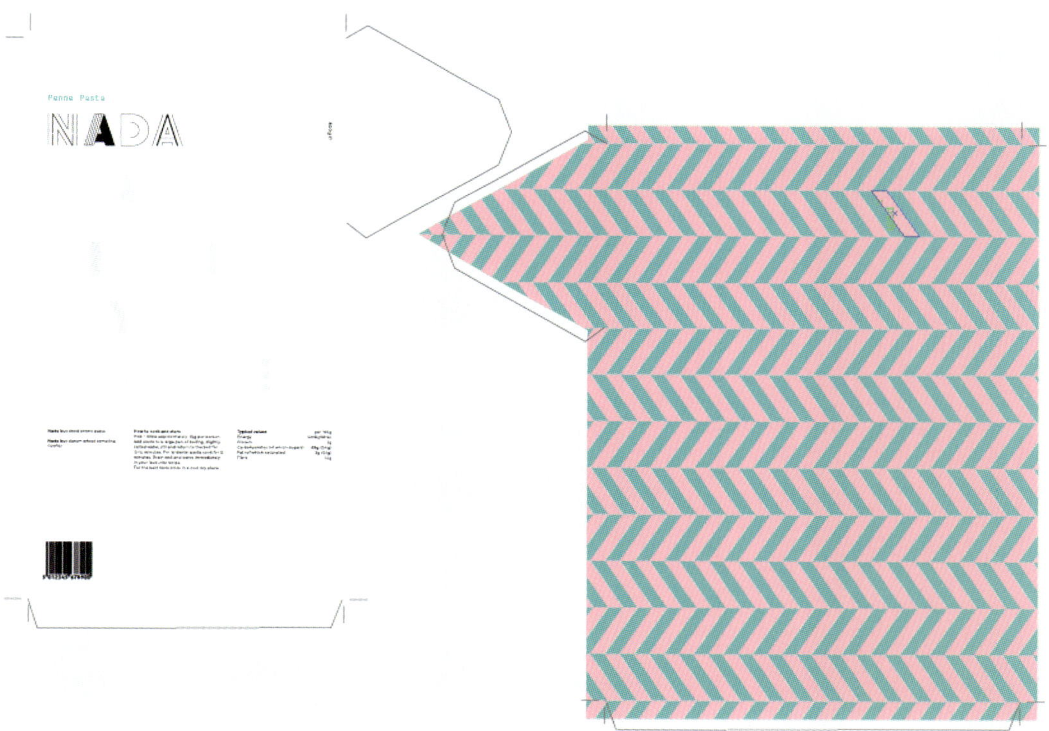
Penne Pasta
NADA

PAPRIKA PACKAGE DESIGN

When I started to design the packaging of paprika I considered several ways because I wanted to avoid the packaging style from the grocery shops and I wanted to make something special.

The emblem has a classy, illustrative style because I think it expresses the tradition of paprika in Hungary, and in the typography I used a ligature which form is similar to the shape of paprika. At the other part of the packaging, I applied a playful method which means I used the details of the logo thus I made a pattern which decorates the packaging. I differentiated the flavours by the help of two distinct colours - the white and the red - so this way the red package is the spicy one and the white is the no spicy.

Designer:
Eszter Varga

Photography:
Mihány Radványi

paprika
paprika
CSEMEGE

MAGYAR FŰSZER
paprika
CSEMEGE

paprika
rika

MAGYAR FŰSZER
paprika
CSEMEGE
FINOMRA ŐRÖLT
KÜLÖNLEGES PIROSPAPRIKA
200g
HUNGARY

GYAR FŰSZE

LIMA SOY SAUCE "MAIKO"

Strategy

Lima wanted to celebrate their 50 years of experience, as an innovative pioneer in qualitative organic food, with a limited edition of authentic Tamari soy sauce.

Make it adorable and authentic, think creative and go for a strong impact! With this compact briefing we started working. And we were never more convinced of the strong result! We just dressed the elegant bottle as a Japanese geisha in her original costume with the playful form of the cap as an original Japanese hat. The design embodies the authenticity of the pure Japanese culture in a harmonious and creative way.

Moreover, we succeeded to make a perfect match between design and shape. Frivoulous into every detail, styled with a spontaneous smile, the design fits like a glove onto the well curved bottle. We were all in love with her, so we called her 'Maiko', as a tribute to all geisha's.

A great story on shelf, ownable and incomparable as a humble hostess on your dining table, ready to serve all your favorite dishes. And, she's so lovely when she smiles!

Design Agency:
QUATRE MAINS, Belgium

Designer:
Patrick De Grande, Kobe De Keyser, Vicky Acke, Thomas Minnaert

Client:
Lima

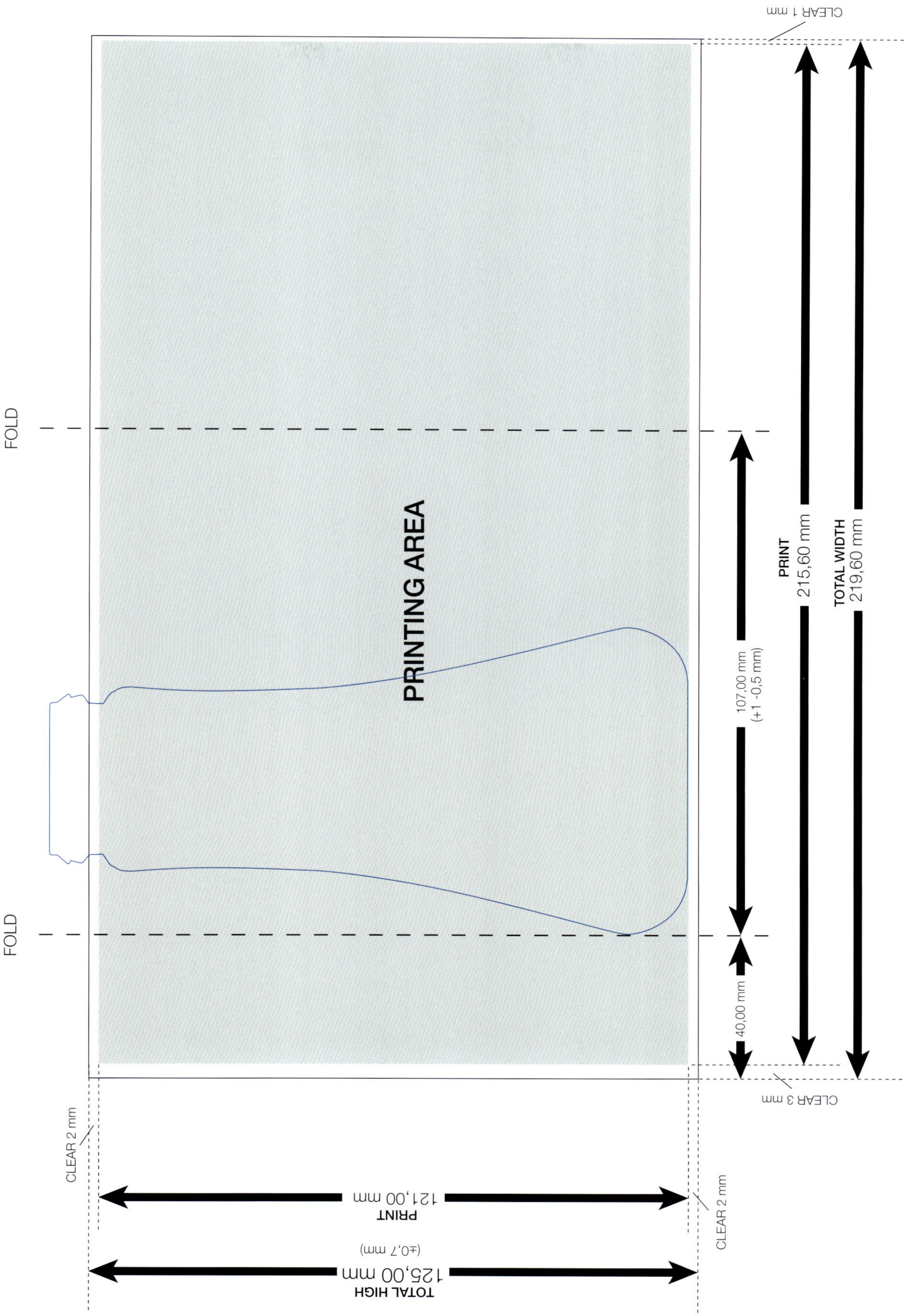
CLEAR 1 mm
FOLD
FOLD
PRINTING AREA
107,00 mm
(+1 -0,5 mm)
40,00 mm
PRINT
215,60 mm
TOTAL WIDTH
219,60 mm
CLEAR 3 mm
CLEAR 2 mm
CLEAR 2 mm
PRINT
121,00 mm
TOTAL HIGH
125,00 mm
(±0,7 mm)

RAVE REVIEW BOTTLE & PACKAGING

Rave Review! Original Culinary Spirits are the world's first collection of spirits specifically crafted for cooking. Inspired by this innovation, the shapely silhouette of the Rave Review! Bottle reflects the uniqueness of the product. The forward-looking design is an elegant companion for food preparation, possessing a sophistication that is chef-worthy.

The flaring form of the bottle is designed to stand proudly and securely in a busy kitchen. The tapered neck helps control the amount of culinary blended spirit needed in a recipe. The bottle also has a unique, custom synthetic cork which is inspired by a chef's hat that when removed can be laid on the counter with its end raised while not rolling. The graphics are bold and color-coded to first attract and then educate customers while also offering minimal cues that exhibit Rave Review! As a spirit for cooking.

Design Agency:
Karim Rashid

Designer:
Karim Rashid

Client:
Rave Review

Photography:
Karim Rashid

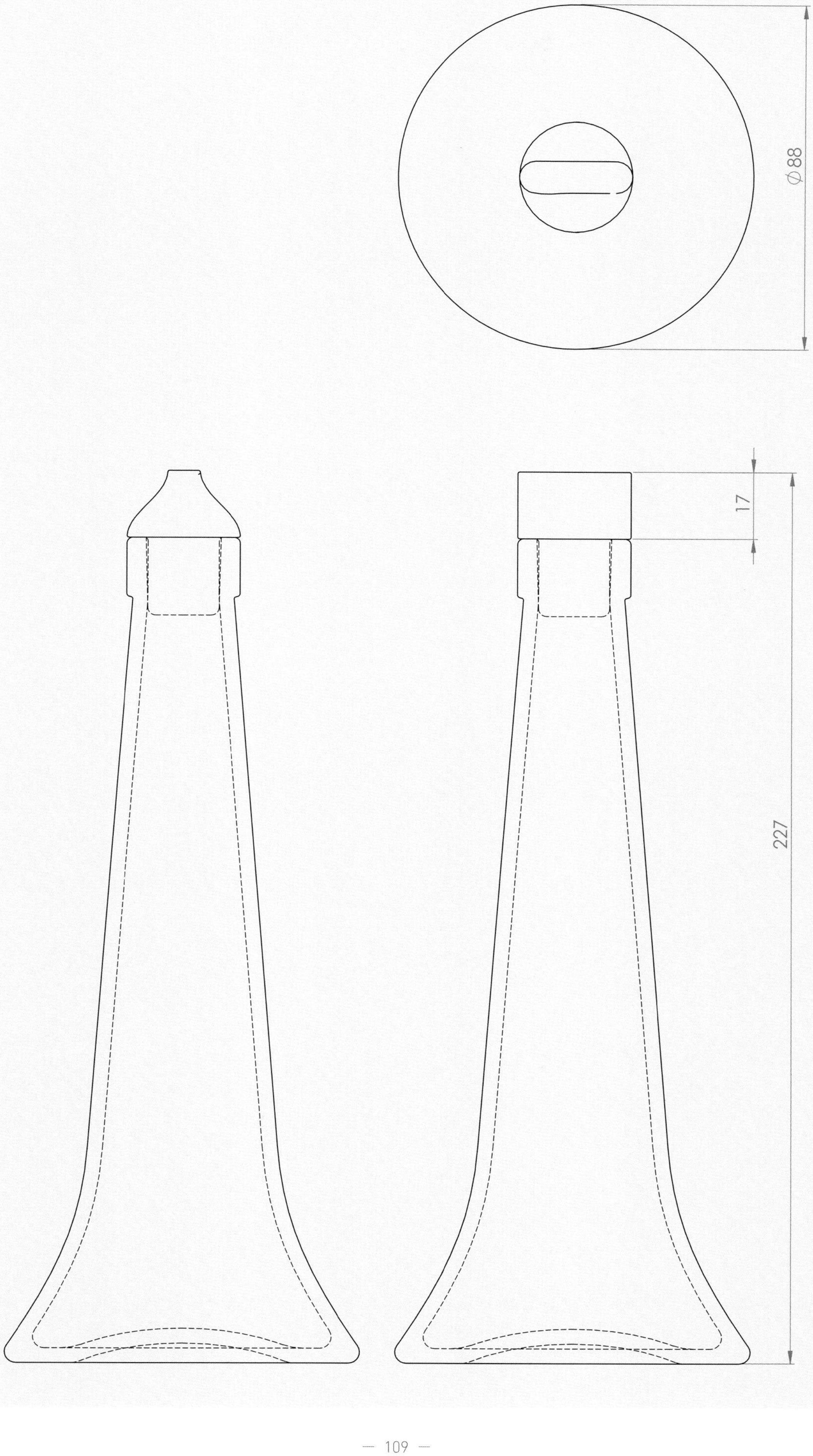
Ø88
17
227

DILLII, COOKIES PACKAGING

ESAD/CR (University of Arts and Design, Caldas da Rainha Portugal)
University Project

Dillii with the motto Dillicious&Dillirious is a project I did during my graduation in ESAD Caldas da Rainha Portugal, I had to make the packaging and branding identity for a fictional organic cookies brand, the brief wasn't that restrictive so I tried to have the most fun with it, since I have a special love for illustration I made it very visual but at the same time with the concerned in conciliating the interact part with the illustrations, so I ended up making it in a way that when opened the illustrations change in a delirious way, showing new narratives and the sensations in tasting the cookies (ice cream man melting for hot spicy cookies, and astronaut space fresh for the mint cookies) making it fun to interact, just like opening a wrapped gift releasing all the surprise and joy in somebody's smile.

Designer:
Rúben Rodrigues

Dillii
CHOCOLATE PICANTE
Biscoito de chocolate picante
Ingredientes:
manteiga sem sal (30%), chocolate, açúcar amarelo (20%), açúcar granulado, extracto de baunilha (15%), ovo, farinha, cacau em pó, canela, tempero Mexicano de pimento, pimento cayenne, sal, gotas de chocolate.
Valores nutricionais medios* a/100g
Valor energetico 2102 kj/ 502 kcal
Proteinas 6,7g
Hidratos de carbonodos 61,0g
dosquais açucares 36,0g
Lipidos 25,0g
dos quais saturados 14,0g
Fibras alimentares 3,3g
Sódio 0,27g
PRODUTO BIOLÓGICO
CHOCOLATE PICANTE
Galletas
2.5€
200g
IRANTES & ICIOSAS
*Valores diários de referência para adultos.
Conservar em local fresco e seco, Depois de aberta, conservar a embalagem fechada de forma a manter todas as características do produto.
Embalagem de papel reciclado amiga do ambiente. Produto biológico certificado
200g
10 biscoitos
um produto:
Dillii
ECOPONTO AZUL
Para mais informação contactar o apartado:
DelRubro & Familia Rua do Cartaxo nº8
3020-944, Vil de Matos Coimbra.
ruro.rush@gmail.com
Biscoitos • Cookies
Dillii
RUBRO

Design Agency:
Inter Group

Art Director:
Gleb Gafl

Illustration:
INORAMA illustrators

Client:
the Chocolate Factory ЯR,
Yaroslavl, Russia

DEVELOPMENT OF PACKAGING DESIGN CHOCOLATE ЯR. 2015

Illustrations for chocolate

For the package, INORAMA illustrators drew the nice almonds and hazelnuts.

SUGAR COATED

Sugar Coated is a project we truly fell in love with. We were given the opportunity to create the whole brand starting from the logo all the way to the booth design. Sugar Coated is a Kuwaiti business that creates delicious chocolate dipped strawberries; we wanted to capture the essence of the taste and nostalgic themes throughout the brand, packaging, photography, website and booth.

Sugar Coated
OH MY!

Design Agency:
Studio AIO

Designer:
Moodhi Alghanim & Ahmed Alrefaie

Photography:
Studio AIO (strawberry pictures)

THREE
DOZEN
GOODNESS
Sugar Coated
OH MY!
94466302/4
SUGARCOATED
SUGARCOATED.COM
THREE
DOZEN
GOODNESS
Sugar Coated
OH MY!

OH MY!
IT'S SO
TASTY
OH MY!
IT'S SO
TASTY

Sugar Coated
OH MY!
OH MY!
IT'S
HAZELNUT

WE ARE
MAKING
A FRESH
BATCH
SOON
OH MY!
WE ARE
SOLD OUT
94486392/4
SUGARCOATED
SUGARCOATED.COM
Sugar Coated
OH MY!

OH MY!
YOU'RE
SO GOOD
OH MY!
YOU'RE
SO GOOD

OH MY!
YOU'RE
SO GOOD
OH MY!
YOU'RE
SO GOOD
OH MY! IT'S TIME TO

OH MY!
I'M IN
HEAVEN
Sugar Coated
OH MY!
OH MY!
I'M IN
HEAVEN

Sugar Coated
OH MY!
OH MY! IT'S SO TASTY
OH MY! IT'S SO TASTY
OH MY! IT'S SO TASTY
94466302/4
SUGARCOATED
SUGARCOATED.COM

Sugar Coated
OH MY!
OH MY! YOU'RE SO GOOD
OH MY! YOU'RE SO GOOD
OH MY! YOU'RE SO GOOD
94466302/4
SUGARCOATED
SUGARCOATED.COM

TAIWAN TRAVEL SERIES

Based on the theme of "visiting tourist attractions in Taiwan", the brand new TK Food Gift Box uses retrospective and smart 8-bit illustrations to form the images of cubic, representing piles of tasteful cubic pastry. Easy to carry, the trapezoid-shaped plastic gift box brings you the wonderful childhood memory of sharing the candy in Kuai Kuai Gift Box with your family, relatives and friends.

Design Agency:
Victor Branding Design Crop.

Client:
TK food

T.K FOOD
Baked Wheat Cookies
Taiwan Native Flavor
EXCELLENT BAKED COOKIES

A WILD ROSE FROM TINE

Tine has collected the best taste experiences from regions with a long and rich culture. The product range consists of five products. Milk from Valdres, Cream from Valdres, Cottage cheese from Trysil, Light Sour from the North and Butter from Setesdal. The painted rose decorates the products with bold and free brushwork. Colors are exhilarating and fresh against white background. These are products created from local conditions and traditions, which is reflected in the visual.

Design Agency:
Tangram Design

Illustrator:
Arild Sæther

TINE
SKJØROST
FRA TRYSIL
FERSKOST AV SYRNET MELK

INGREDIENSER:
Pasteurisert skummet melk
KJØLEVARE: 0 °C – 4 °C.
BEST FØR: Se påført dato
NÆRINGSINNHOLD: 100 g vare gir ca.: Energi 448 kJ (97 kcal), fett 0,6 g, - hvorav mettede fettsyrer 0,4 g, karbohydrat 2 g, -hvorav sukkerarter 2 g, protein 21 g, salt 0,1 g*, kalsium 830 mg (104 %), fosfor 340 mg (49 %).
' av referanseverdien
*Saltinnholdet er beregnet fra naturlig innhold av natrium i råvarene.
Nettoinnhold: 160 g
TINE SA
Pb 25, 0051 Oslo
TINE Forbrukersenter 800 35 999 www.tine.no
NYT NORGE
NORGE M 147 EFTA
7 038010 052569
TINE
SKJØROST
FRA TRYSIL
FERSKOST AV SYRNET MELK
TINE
SKJØROST
FRA TRYSIL
HELT SIDEN VIKINGETIDEN HAR DET BLITT LAGET SKJØROST I TRYSIL. DEN ER LAGET MED SYRNET SKUMMET MELK, NOE SOM GIR EN GOD SMAK OG ET NATURLIG HØYT INNHOLD AV PROTEIN. DEN MAGRE OSTEN PASSER PERFEKT I SALATER ELLER WRAPS. LES MER PÅ TINE.NO

TINE
LETTRØMME
FRA NORD
MELK FRA GEIT OG KU

INGREDIENSER:
Geitefløte, skummet melk og melkesyrekultur. Pasteurisert og homogenisert før syrning.
KJØLEVARE 0 °C – 4 °C.
BEST FØR: Se påført dato
NÆRINGSINNHOLD: 100 g vare gir ca.: energi 842 kJ (204 kcal), fett 20 g, -hvorav mettede fettsyrer 13 g, karbohydrat 3,4 g, -hvorav sukkerarter 3,4 g, protein 2,7 g, salt 0,1 g*, vitamin A 150 µg (19 %'), kalsium 120 mg (15 %') ' av referanseverdien
*Saltinnholdet er beregnet fra naturlig innhold av natrium i råvarene.
Nettoinnhold: 280 g
TINE Forbrukersenter 800 35 999 www.tine.no
NYT NORGE
NORGE M 350 EFTA
7 038010 052590
TINE
LETTRØMME
FRA NORD
MELK FRA GEIT OG KU
TINE
LETTRØMME
FRA NORD
FOR Å LAGE EN RØMME SOM SMAKER EKSTRA GODT, MÅ MAN BRUKE EKSTRA GODE RÅVARER. VI HAR BRUKT BÅDE GEITEFLØTE OG KUMELK FRA NORD-NORGE, NOE SOM GIR EN FRISK OG RUND SMAK SOM PASSER TIL DET MESTE. LES MER PÅ TINE.NO

TINE
STØLSMJØLK
FRÅ VALDRES
EI SMAKSRIK, LETT TINEMJØLK
SPESIALITET
UNIK SMAK

TINE
STØLSFLØYTE
FRÅ VALDRES
KREMFLØYTE FRÅ KYR PÅ SOMMARBEITE

TINE
SETESDALSMØR
SYRNA ØKOLOGISK FLØYTE KJERNA
PÅ TRADISJONELT VIS
I SETESDAL HAR DEI KINNA SMØR PÅ TRADISJONELT VIS I UMINNELEGE TIDER.

Kartongen kan resirkulerast med kork
TINE
STØLSMJØLK
FRÅ VALDRES
EI SMAKSRIK, LETT TINEMJØLK
SPESIALITET UNIK SMAK
Stølsmjølk frå Valdres er som eit kjærleiksbrev frå fjellheimen. Her, 900 meter over havet, lètdei lokale mjølkebøndene kyrne gå på beite gjennom heile sommaren. Ingen andre stader i norden har fleire aktive stølsbruk enn Valdres. Mjølka blir foredla med kunnskap og kjærleiktil ei fyldig rik lettmjølk av ypparste kvalitet.
BLÅ HIMMEL, GRØNKLEDDE VIDDER, STILLE OG FRISK LUFT. NÅR FJELLHEIMEN I VALDRES BYR PÅ SITT ALLER BESTE, FÅR MJØLKEKYRNE ALT DEI KAN DRØYME OM. TAKKA VERE MEIR ENN 250 SOMMARSTØLAR I FULL DRIFT HAR VALDRES FRAMLEIS EIT RIKT OG LEVANDE KULTURLANDSKAP.
NORDMENN ELSKAR Å GÅ I FJELLET. DET GJER KYRNE OG. DEI STOLTE MJØLKEBØNDER I VALDRES FØLGT KYRNE SINE TIL FJELLS KVAR SOMMAR I MANGE HUNDRE ÅR. PÅ STØLEN VANDRAR DYRA FRITT MELLOM GRØNE URTER, SMAKSRIK LYNG OG SAFTIG FJELLGRAS OG GIR DEG NOREGS ALLER FINASTE MJØLK.
LETTMJØLK
Homogenisert og pasteurisert
Ingrediensar: Lettmjølk
NÆRINGSINNHALD
100 g vare gir ca.:
energi 201 kJ (48 kcal)
feitt 1,8 g
av dette metta feittsyrer 1,1 g
karbohydrat 4,5 g
av dette sukkerartar 4,5 g
protein 3,4 g
salt* 0,1 g
riboflavin 0,15 mg (11 %')
vitamin B12 0,5 µg (20 %')
kalium 171 mg (9 %')
kalsium 134 mg (17 %')
fosfor 115 mg (17 %')
jod 19 µg (13 %')
' av referanseverdien
* Saltinnhaldet er berekna frå det naturlege innhaldet av natrium i mjølka.
Kjølevare: 0 °C – 4 °C. Best før: Sjå påførd dato
Nettoinnhald: 1 liter
7 038010 044717
TINE Forbrukersenter
800 35 999
www.tine.no
FSC MIX Kartong frå berekraftige kjelder FSC® C014047
NORGE * EFTA
*Sjå autorisasjonsnummer i toppen
Ved kjøp av denne kartongen bidrar du til å ta vare på skogane i verda.
NYT NORGE
TINE
TINE SA
Pb. 25, 0051 Oslo

TINE
SETESDALSMØR
SYRNA ØKOLOGISK FLØYTE KJERNA PÅ TRADISJONELT VIS
I SETESDAL HAR DEI KINNA SMØR PÅ TRADISJONELT VIS I UMINNELEGE TIDER. SURNA ØKOLOGISK FLØYTE GIR EIT SMAKSRIKT SMØR, SOM PASSAR PERFEKT TIL TRADISJONSMAT ELLER RETT OG SLETT PÅ BRØDSKIVA. LES MEIR PÅ TINE.NO

Pure-Pak
TINE
STØLSFLØYTE
FRÅ VALDRES
KREMFLØYTE FRÅ KYR PÅ SOMMARBEITE
800 35 999
TINE

LUNG-CHING EGG ROLL

Based on a playful game, this limited edition egg roll packaging design has a unique structure. Inspired by circus elements, illustrator BaNAna created two cute little characters – Mr. Peanut and Ms. Almond – with bold colours. Two main colours blue and yellow were applied to differentiate flavours. Its beautiful graphic design and unique handle make it the perfect gift to carry home.

Design Agency:
Victor Branding Design Crop.

Designer:
Wan Ju Lee

Illustrator:
BaNAna

Client:
Lung-Ching

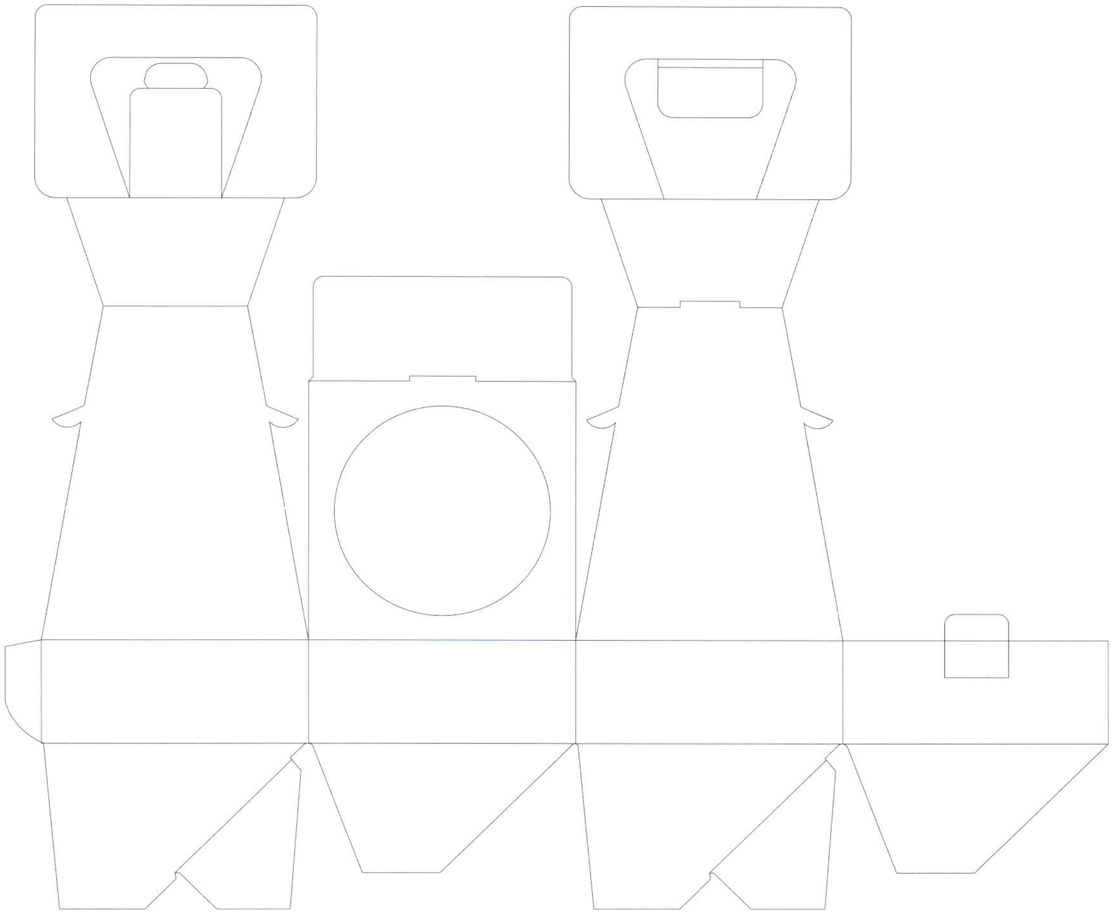

LUNGCHING
PEANUTS
EGG ROLLS
花生蛋捲 crispy and yummy !!!

LE PETIT

The challenge: to design a completely different packaging for the frozen dessert segment in Brazil. All packages have 2 more special PANTONE colors. All chosen colors would have to form a Line in the freezers and distinguishable from each other. The effect chosen for all paper boxes were matte lamination.

I established a communication structure: brand plus illustration with photographic elements, CGI and hand painting by the artist Gabriel Bueno. The color margin breathes make the illustration popup in the package.

First were created the Petit Gateau (Mountain Lava Cake) line with 4 products, and then pies composed by 3 different pies, and finally the brownie snack that is not a product line.

Design Agency:
TYPED STUDIO

Designer/ Creative director:
Carlos Nogueira

Illustrator:
Gabriel Bueno

Client:
Sweet Petit

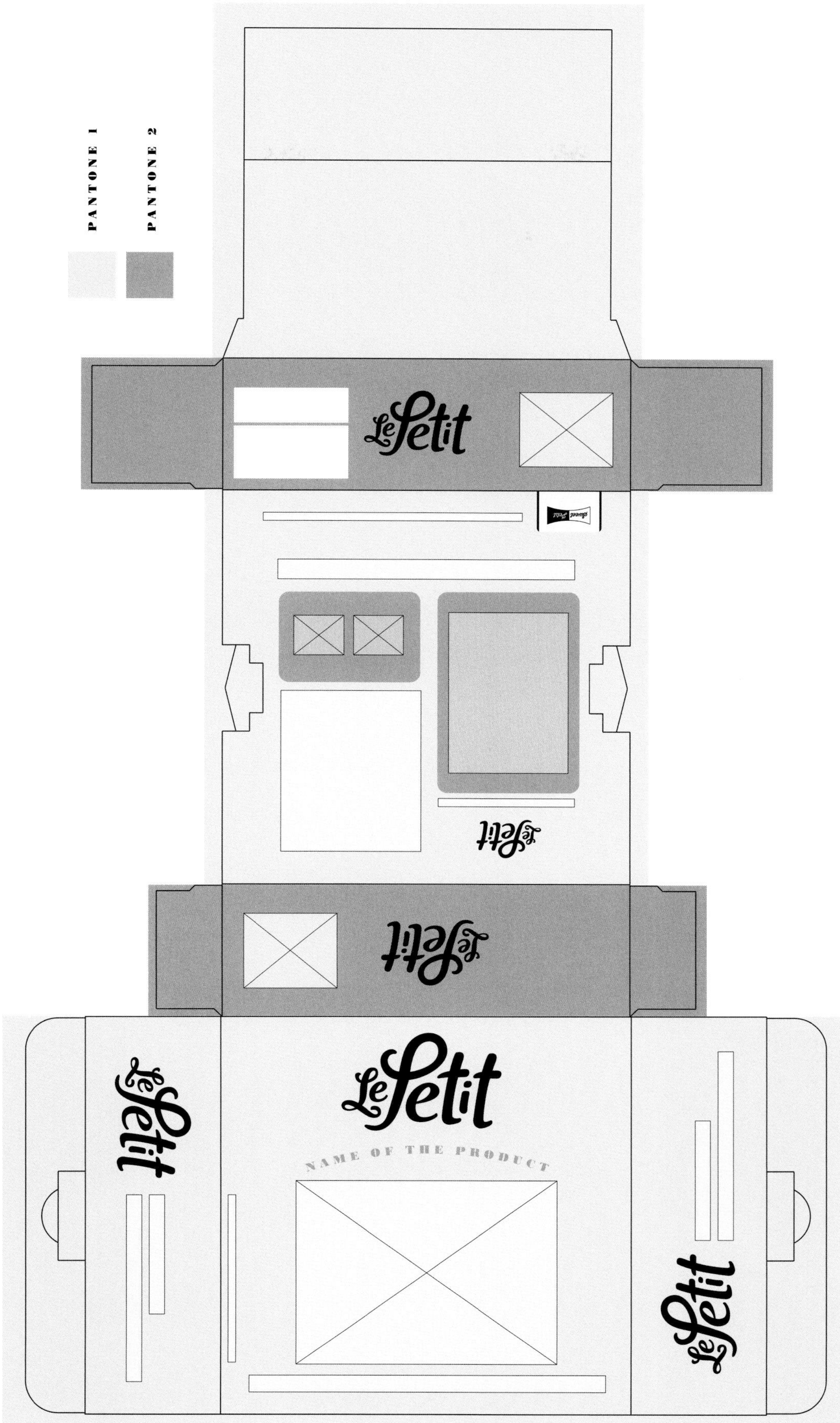
PANTONE 1
PANTONE 2
Le Petit
NAME OF THE PRODUCT

FOOD FOR THOUGHT

Some books are just books and some are "Food for Thought". The limited edition series includes 3 books ("Breakfast at Tiffany's", "Naked Lunch" and "Dinner at the Homesick Restaurant") in a special packaging. The 3 books with edible names contain inspiring novels that encourage profound thinking.

With the help of illustration the packaging sheds light on the contents without revealing the plot. By looking at the packaging you can find out the "daily value" of humor and irony that the book contains, the number of pages that make the perfect "serving size" and the ingredients of the novel. The product group on the packaging gives the chance to learn a few details about the main character at a glance.

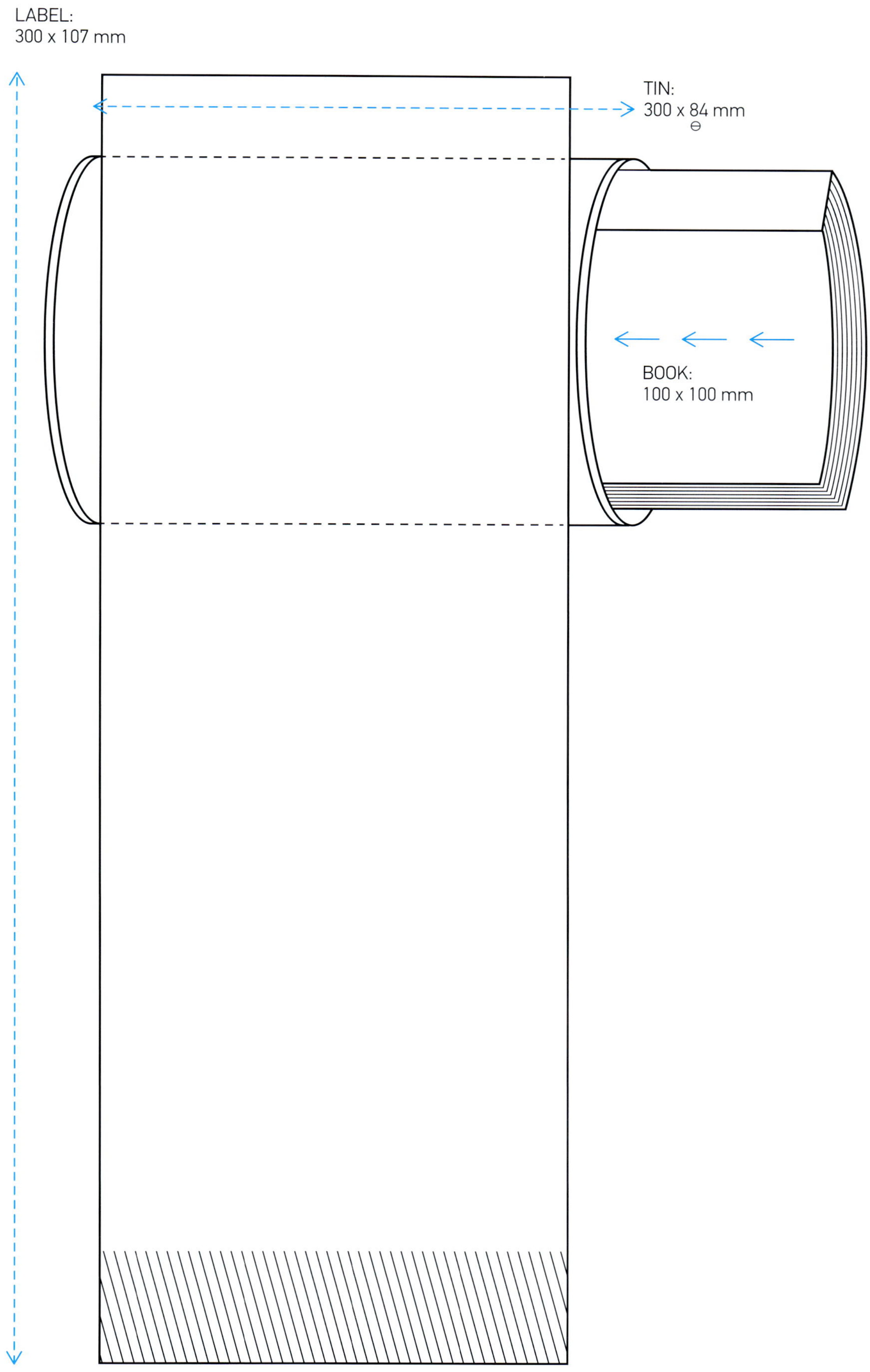
LABEL:
300 x 107 mm
TIN:
300 x 84 mm
⊖
BOOK:
100 x 100 mm

Designer:
Maria Mordvintseva-Keeler

Illustrator:
Maria Mordvintseva-Keeler

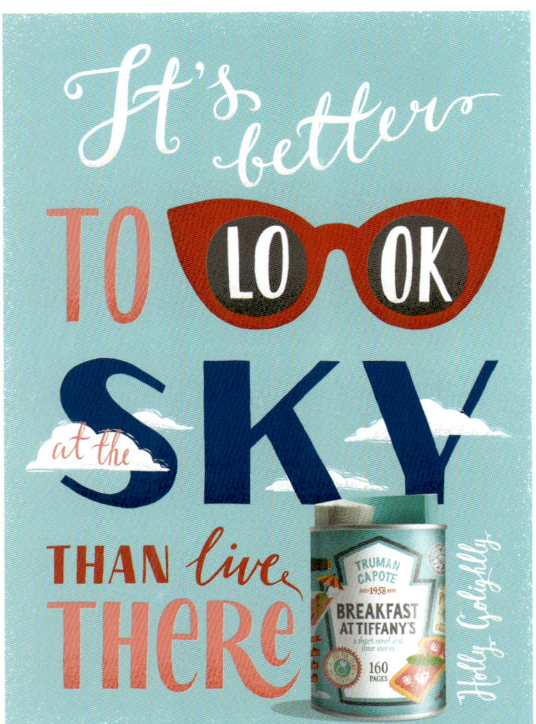

NUTRITION FACTS:
SERVING SIZE 30 PAGES
SERVINGS PER CONTAINER 10 (300 PAGES)
% DAILY VALUE
WISDOM-57
AMUSEMENT-35
+A DASH OF HEARTBREAK
Specials
PEARL TULL
CODY & EZRA
Dr. Jenny Tull
RUTH THE STOLEN BRIDE
CRAB CASSEROLE
ACCI DENTS happen
MILK LACED WITH Honey
SECRET INGREDIENT
Scarlatti's RESTAURANT
ANNE TYLER
ESTD 1982 ESTD
DINNER
AT THE HOMESICK RESTAURANT
a novel
300 PAGES
100 PER CENT CONSOLING
NO ARTIFICIAL CHARACTERS
THOUGHT PROVOKING
GOOD SOURCE OF REGRET
Greensleeves
Orioles
LE GODIVEAU DE POISSON
It's not
A TASTE OF HONEY

THE LACUNA
BARBARA KINGSOLVER
?
What is Art
L
ANNE TYLER
ESTD 1982 ESTD
DINNER
AT THE HOMESICK RESTAURANT
a novel
300 PAGES
SECRET INGREDIENT
ROOM
NATIONAL BESTSELLER
EMMA DONOGHUE
WILLIAM BOYD
armadillo
WHATEVER YOU THINK, THINK THE OPPOSITE
Paul Arden
SHAKESPEARE
ŒUVRES COMPLÈTES
I
POÈMES
DRAMES HISTORIQUES
COMÉDIES I
BIBLIOTHÈQUE DE LA PLÉIADE
nrf

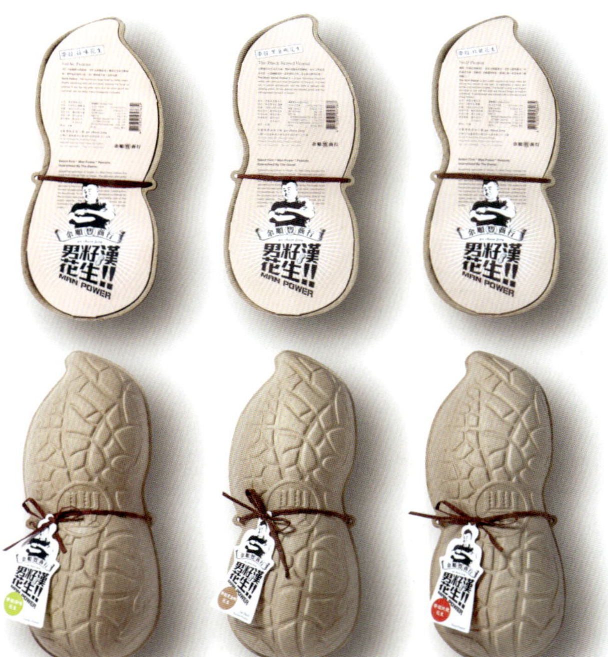

MAN POWER! PEANUTS

Can't help having it one after another. Peanuts with shells are firm as machos. You can simply use two fingers to open the pod and enjoy the crispy texture and luscious taste of the kernel. The package restores with local unsophisticated feeling. The design is a half peanut in molded pulp. The visual image is expressed in humor approach, through the affability and passion which are naturally shown by the third generation of Yu Shun Feng, Mr. Yu Zhao Feng, his frankness and boldness as a true man once again brings a fresh interpretation of peanuts.

Design Agency:
Victor Branding Design Crop.

Designer:
Jui Cheng Liu

Client:
Yu Shuen Feng

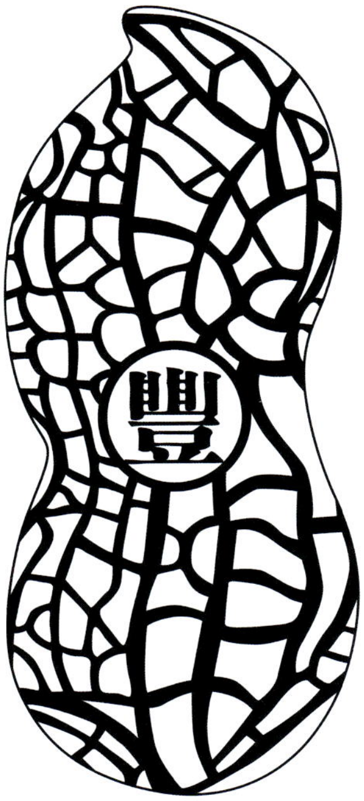
豐

余順豐商行
yu shuen feng
男籽漢
花生!!
MAN POWER

豐
余順豐商行
yu shuen feng
男籽漢
花生!!
MAN POWER
帶殼玖號花生
No.9 Peanut

帶殼.玖號花生
No.9 Peanut
The No.9 Peanut is also called couple nuts since there are always two kernels in one pod. It represents a caring and tender love between couples. The kernel is long oval-shaped and large in size with thin shell and the pod shape is medium cylindrical. It tastes sweet and smooth with a nice aroma and crispy texture.
營養標示 Nutrition Facts
Serving Size 100g
熱量 Calories 579Kcal 大卡
蛋白質 Protein 32g 公克
脂肪 Fat 40g 公克
飽和脂肪 Saturated Fat 8.35g 公克
反式脂肪 Trans Fat 0g 公克
碳水化合物 Carbohydrate 23.5g 公克
鈉 Na 663mg 毫克
4 712305 957465
yu shuen feng
No.20-28, HouDi Rd., Niaosong Village,
Dongshih Township, Chiayi County 614,Taiwan
Tel: 05 3660605 Fax: 05 3660886
余順豐商行
Select Firm " Man Power " Peanuts.
Guaranteed By The Owner.
余順豐商行
yu shuen feng
男籽漢
花生!!
MAN POWER

DE RUYTER SPECIALS

Even better chocolate; Bringing the true cacao feel to your sandwich. The creamy dark background and the golden foil embossing emphasizes the high quality.

Design Agency:
Mountain Design

Designer:
Maurice Ajanaku

Account:
Judith de Kooning

Client:
H.J. Heinz

Photography:
Four Photography

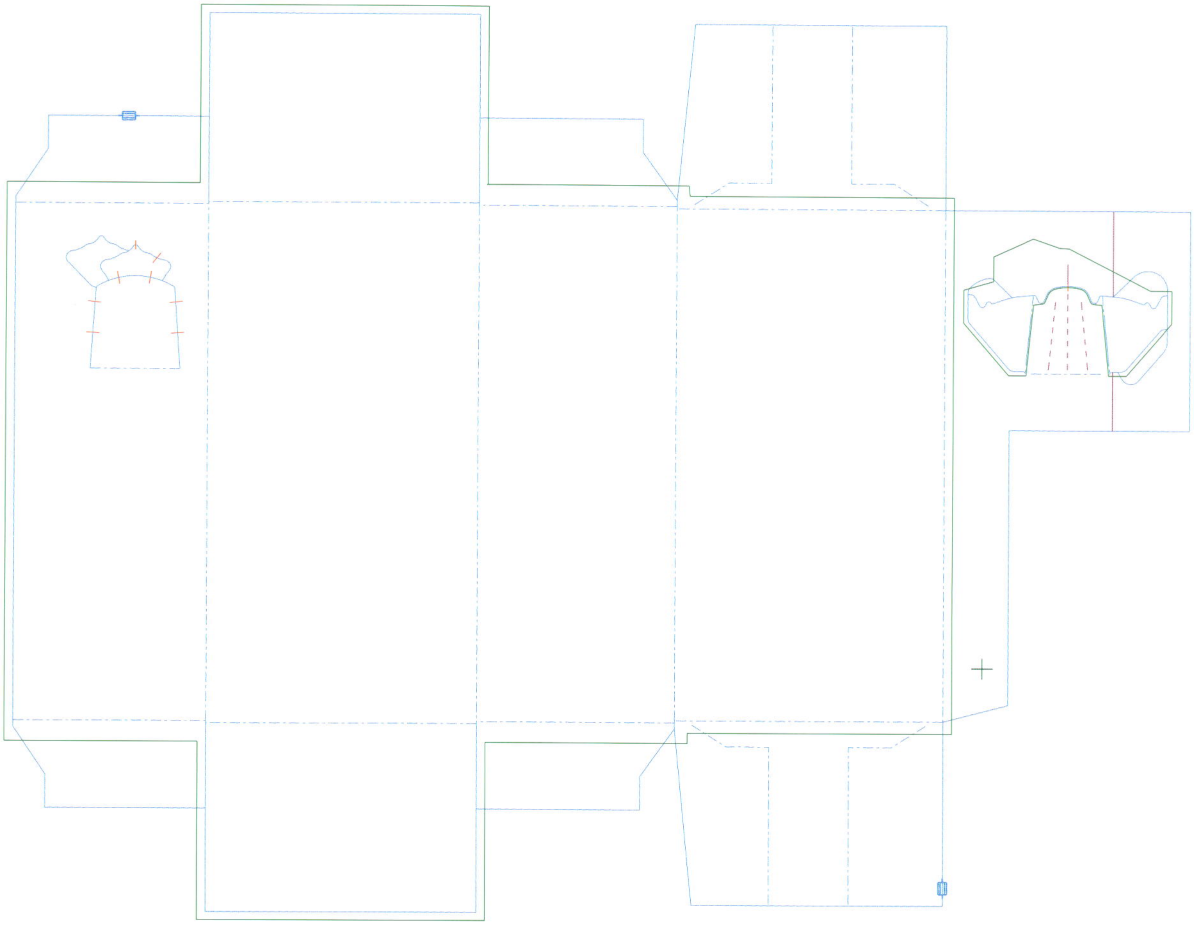

De Ruijter
Specials
Puur-Koffie
verwijderen
Consumentenservice:
Wij willen graag dat iedereen blijft genieten van De Ruijter-producten. Heeft u vragen of opmerkingen? Bel of schrijf naar H.J. Heinz B.V., Postbus 217, 3700 AE Zeist. De Ruijter Consumentenservice, t.a.v. De Ruitjer Consumentenservice consumentenservice@deruijter.nl Bel gratis: 0800-0233370 (werkdagen 9.00-17.00 uur). Ook kunt u informatie vinden op
www.deruijter.nl
UTZ Certified cacao
De Ruijter koopt duurzame cacao voor dit product.
Met intense koffiesmaak
De Ruijter
Sinds 1860
UTZ Certified cacao
Specials
Koffie-Puur
20%
meer cacaobestanddelen
De Ruijter
De Ruijter Specials
Chocoladehagel met meer cacao, voor de échte chocoladeliefhebber! In combinatie met ons vakmanschap maakt dit de smaak van De Ruijter Specials nóg intenser. Pure verwennerij voor op de boterham.
Deze chocoladehagel bevat 20% meer cacaobestanddelen dan De Ruijter Chocoladehagel Puur.
Chocoladehagelslag Puur met Koffiesmaak.
Ten minste 39% cacaobestanddelen.
Ingrediënten:
Suiker, magere cacaopoeder, cacaoboter, emulgator (sojalecithine), natuurlijk aroma.
Kan sporen bevatten van melk.
Ten minste houdbaar tot einde: zie onderzijde verpakking. Koel en droog bewaren.
220g ℮
Voedingswaarde
Per 100gram | Per boterham van 20g
Energie in kJ/kcal 1820 kJ 430 kcal | 365 kJ 86 kcal
Vet 17 g | 3,4 g
waarvan verzadigd 10 g | 2,0 g
Koolhydraten 61 g | 12 g
waarvan suikers 58 g | 12 g
Eiwitten 6,1 g | 1,2 g
Zout <0,1 g | <0,1 g
Met intense koffiesmaak
De Ruijter
Sinds 1860
UTZ Certified cacao
Specials
Koffie-Puur
20%
meer cacaobestanddelen

PLUK KRUIDJE

As herbal as it gets; Pluk! is an innovative concept of Herbal Food For Health. Fresh herbs and natural flavours. Using grandma as an icon on the pack. An humorous and optimistic style provides a modern touch to it all.

Design Agency:
Mountain Design

Designer:
Jon Sonneveld

Account:
Esther Kloosterman

Client:
Herbal Health Food BV

Illustrator:
Jorien Doorn

vers - smaak - gezond
- Melange van bijzondere theekruiden
- Natuurlijk, gezond en zonder cafeïne
- Voor 3x 'n potje kruidenthee
- Theeglas kan 2 x gevuld worden
- Water 80-90°C, op smaak na 3-4 min.
INHOUD IN GR:
TGT (MITS KOEL BEWAARD):
3x Potje thee
morgen
fris
marokkaanse munt,
onsterfelijkheidskruid
en appelmunt
de gezonde
smaakmaker
www.plukkruidje.nl
Pluk!
kruidje

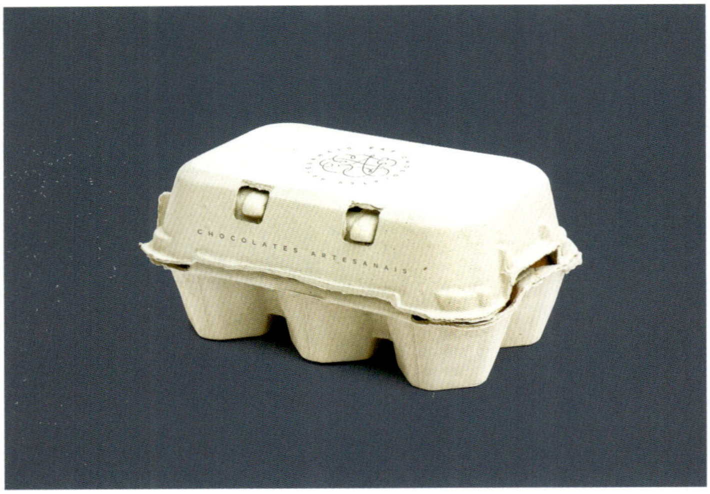

EAT CHOCOLATES

A melted chocolate wire was the inspiration for the development of EAT Chocolates Artesanais Visual Identity. His light colors reflect the subtle and unique touch of flavor that the product has. The variations in packaging represent care in food preparation and at the same time, a search for the use of simple materials that can be found around us.

Designer:
Pedro Paulino

Client:
Eat Chocolates Artesanais

Photography:
Reinaldo Ortlieb

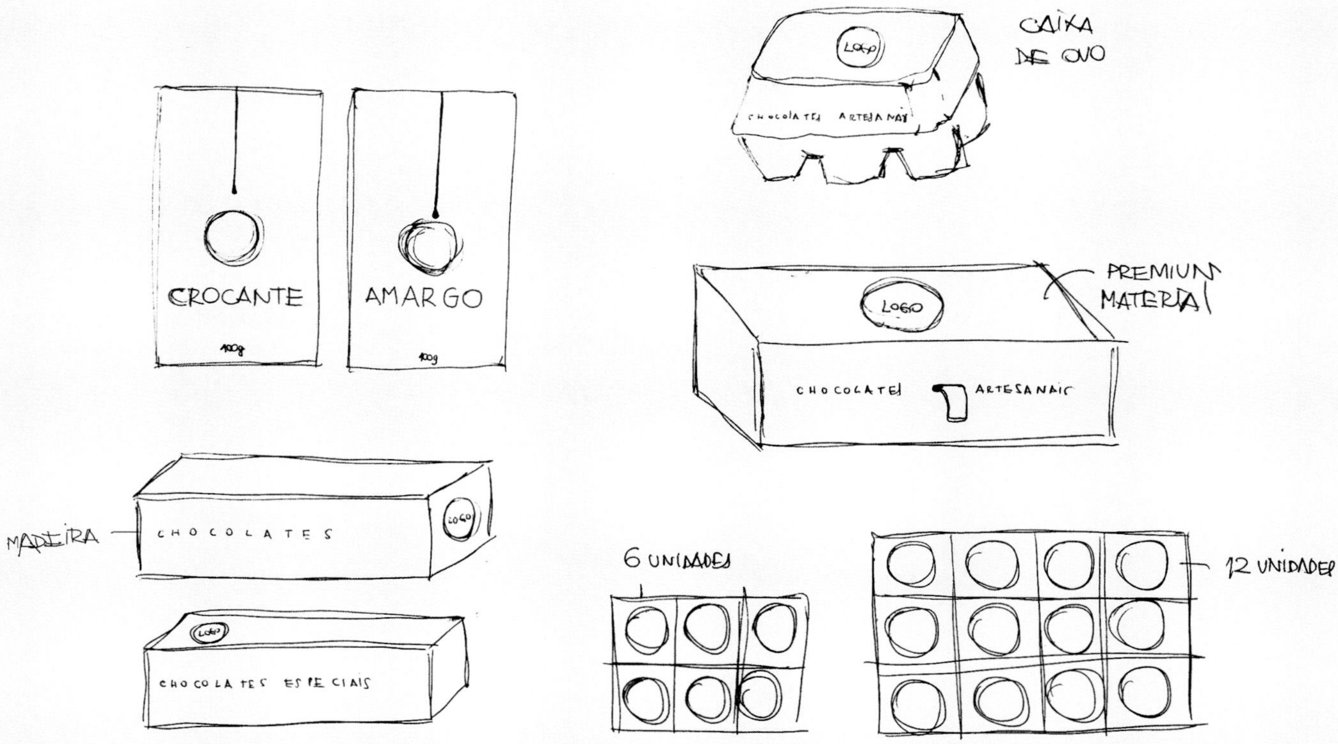

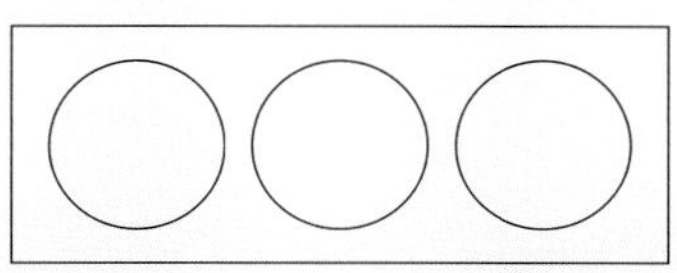

120 x 45 mm

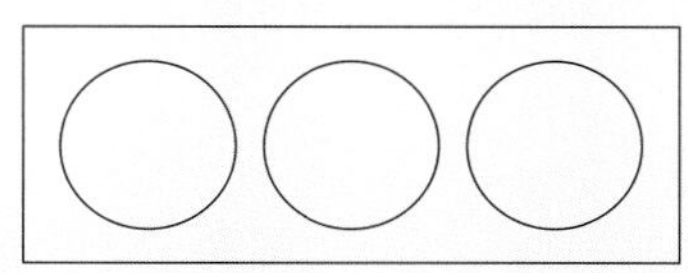

120 x 45 mm

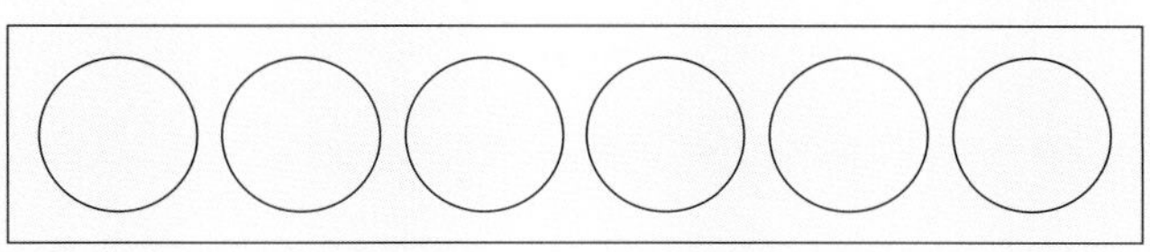

230 x 45 mm

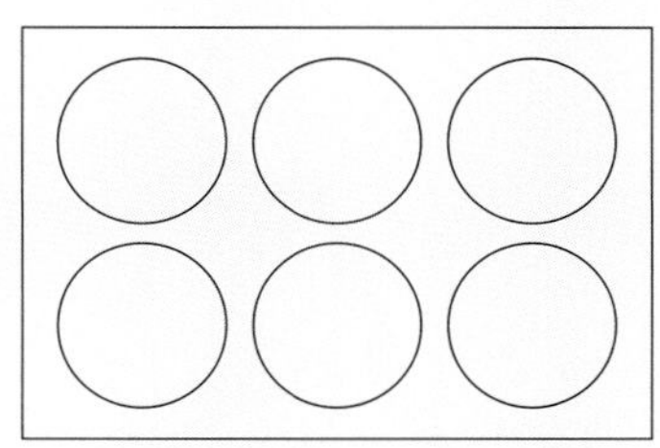

120 x 80 mm

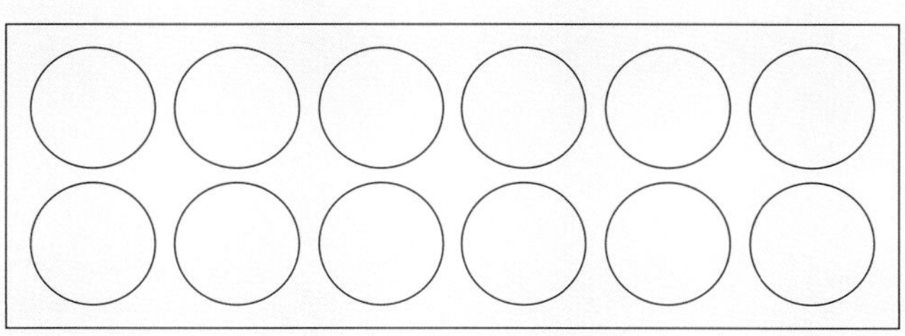

230 x 80 mm

120 x 154 mm

CHOCOLATES ARTESANAIS

ARTESANAIS EAT CHOCOLATES
CHOCOLATES ARTESANAIS

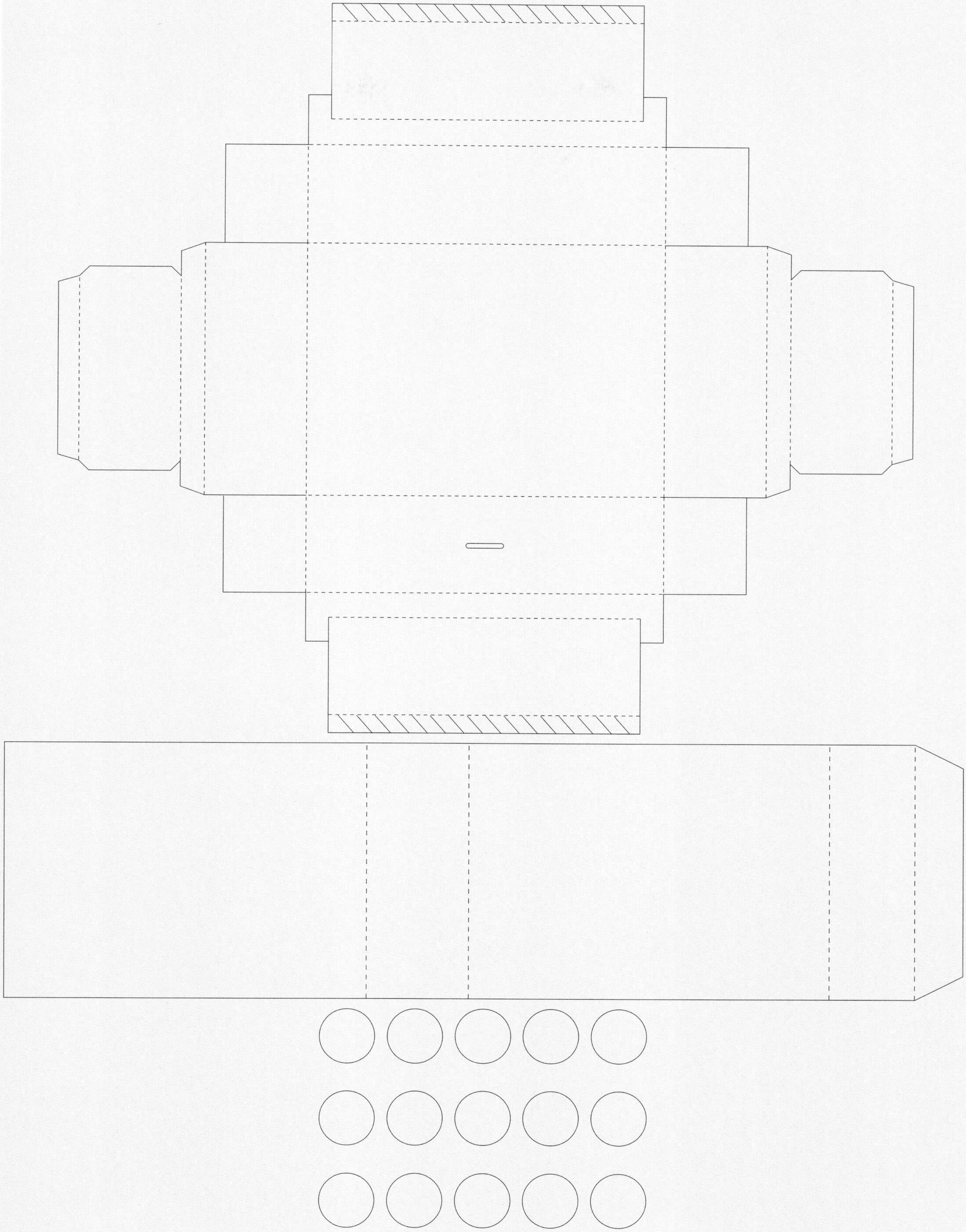

KNEAD BAKERY

Developing a gluten or dairy intolerance requires significant lifestyle changes, often necessitating the preparation of meals from scratch. Few brands offer convenient ready to eat options and those that do, seldom deliver on taste. Knead Bakery set out to change things by offering delicious muffins and cookies good enough for all to enjoy. P&W (Pemberton & Whitefoord) was commissioned to create a packaging solution that was: original, practical and eye catching. The resulting "K Cube" concept gave on shelf presence, product visibility and an ideal platform for essential brand messages.

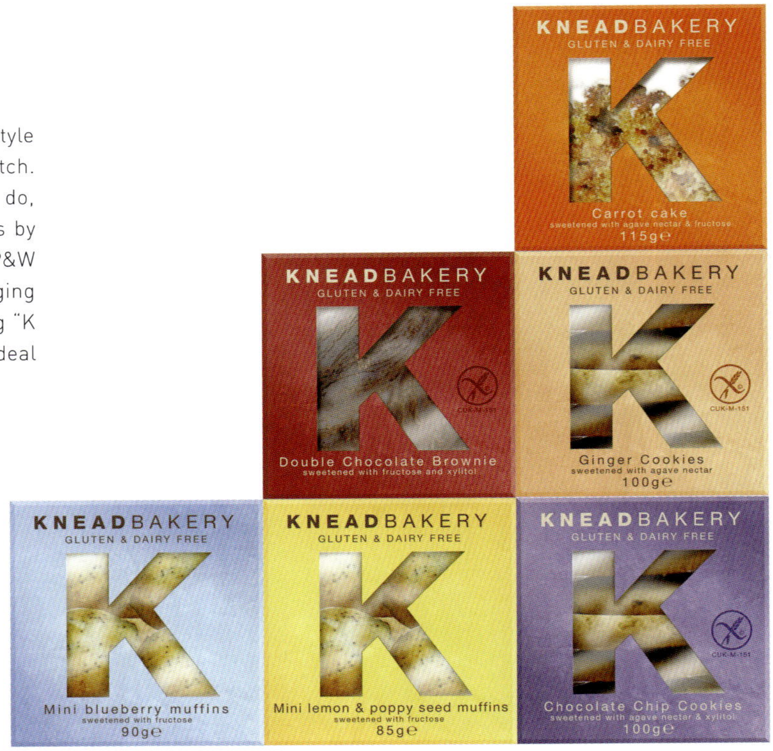

Design Agency:
Pemberton & Whitefoord LLP

Designer:
Jack Bannerman

299.6
21.0
64.0
75.0
47.5
64.5
75.0
65.0
75.0
2
208.1
Ginger
Cookies
pack of 6
No artificial colours or flavourings.
May contain traces of nuts & soya.
Wheat, gluten, dairy, casein,
egg and yeast free.
Sweetened with agave nectar.
Suitable for Vegetarians.
Suitable for Vegans.
Ingredients: Gluten Free Flour (Rice, Potato, Tapioca, Maize & Buckwheat), Agave Nectar, Grape Seed Oil, Brown Rice Syrup, Ground Ginger (1.7%), Lemon Zest, Raising Agent: Bicarbonate of Soda, Salt
Typical values | Per 100g
Energy | 1856kJ 442kcal
Protein | 2.2g
Carbohydrate | 69.5g
Fat | 17.2g
KNEADBAKERY
GLUTEN & DAIRY FREE
K
GUK-M-151
Ginger Cookies
sweetened with agave nectar
100g℮
Ginger Cookies
Incredibly moreish ginger cookies made with agave nectar. Soft, spicy and delicately flavoured with a touch of lemon zest. Baked in small batches, they're lovingly handmade with you in mind.
info@kneadbakery.co.uk
0843 289 5024
Knead Bakery Ltd, PE19 2JL
Produced in the UK
Store in a cool, dry place.
Best Before - see base
5 060265 810016
'oK so are you struggling to find suitable products for your diet in the free from aisle? I was too, which is how Knead Bakery came about. I wanted to cater for people like me. People with not just one, but multiple allergies and intolerances.
Offering something new to the free from market, all of Knead Bakery's products are gluten, dairy and yeast free. We never use cane sugar, artificial colours or preservatives in our products either. So, whether you're coeliac, dairy intolerant or just health conscious, our great range of products is definitely worth a taste. They're delicious!'
Sim
(Founder of Knead Bakery)
www.kneadbakery.co.uk
RECYCLABLE PACKAGING

TESCO SOUP

The new design brings the soups up to date with current trends in the chiller cabinet utilising the tetra pack format.

Based on the idea of the product coming fresh from farm to table, the packs show an abundance of vegetables to promote the 1 of 5 a-day content.

The modern, vibrant, watercolour illustrations bring the concept of freshness to life, along with the holding device which is inspired by gardeners' seed packets, with hand written type, making the packs feel friendly and approachable.

A limited edition range was used to support the core offering and relate to specific seasons – for example the simulated hand knitted jumper illustration eluding to warmth for the winter season.

Design Agency:
Pemberton & Whitefoord (P&W) International Design Consultants

Designer:
Laura Grievson

Illustration:
Mary Woodin

Make sure both ends are opened back.
Pull open pack from diagonally opposite corners.
Open both sides.
Allergy advice:
• Recipe: No nuts. • Ingredients: Cannot guarantee nut free.
• Factory: Before being prepared for manufacture of this product, the equipment was previously used to make products containing nuts. Suitable for vegetarians
Caution:
• Take care when removing and pouring as carton will be hot.
Ingredients: Tomato (71%), Tomato Juice, Tomato Purée, Onion, Garlic Purée, Sugar, Cornflour, Salt, Vegetable Oil, Red Chilli, Black Pepper.
Nutrition:
Typical values. 100g contains: Energy 136kJ/32kcal, Protein 0.8g, Carbohydrate 4.8g (of which sugars 3.2g), Fat 1.1g (of which saturates 0.2g, mono-unsaturates 0.5g, polyunsaturates 0.3g), Fibre 0.6g, Sodium 0.2g, Salt equivalent 0.5g. For guideline daily amounts, please visit: www.tesco.com
Storage: Keep refrigerated.
• Use by: see top of pack.
Additional information:
• We haven't added any artificial preservatives, flavours, colours or hydrogenated fat into this recipe.
DISPOSAL - CARTON - CARD
Widely recycled
600g e
Produced in the U.K. for Tesco Stores Ltd., Cheshunt EN8 9SL, U.K. © Tesco 2011. SC0112C
5 052320 877010
1 of 5 a day
TESCO
PEA & ASPARAGUS
SOUP
Each portion contains
Calories 87 4%
Sugar 8.1g 14%
Fat 4.1g 5%
Saturates 2g 7%
Salt 0.1g 2%
of your guideline daily amount
Suitable for home freezing
PURE-PAK®
Make sure both ends are opened back.
Pull open pack from diagonally opposite corners.
Open both sides.
WHY NOT TRY THESE...
All our soups are created by our team of dedicated chefs using only the best possible ingredients. Every soup is quality tested by Tesco customer ensuring that only the very best recipes make it on to our shelves.
• Shake carton well and open it up fully.
Microwave
750 watt Category D
850 watt Category E
Heat on full power then stir 3 mins 30 secs 3 mins 30 secs
Heat on full power for a further 3 mins 30 secs 2 mins 30 secs
Stand 2 mins 2 mins
Total time 9 mins 8 mins
Hob • Shake carton well and open it up fully. 5–6 mins
• Empty contents into a saucepan.
• Heat gently for 5–6 minutes, stirring occasionally until piping hot
• Do not allow to boil.
Freezing guidelines
• Freeze on day of purchase. Use within one month.
Microwave from frozen
• Open carton fully.
750 watt Category D
850 watt Category E
Heat on full power then stir 6 mins 30 secs 6 mins 30 secs
Heat on full power for a further 5 mins 4 mins 30 secs
Stand 2 mins 2 mins
Total time 13 mins 30 secs 13 mins
• Adjust times according to your particular oven.
• All appliances vary, these are guidelines only.
• Check food is piping hot throughout before serving.
• If not, reheat for a short time.

DROP HONEY PACKAGE

Drop honey package is both: unique, exclusive shape and functional, easy for production jar. It has pressed flowing down drop silhouette on the edge which may remind honey drop. The label highlights this feature.

The shape of the bottle is good for logistics and has the enough room for big label with much info. The form is suitable for mass production.

The package has been designed for Simplicity Company.

Designer:
Igor Solovyov

Client:
Tan Ren Tan

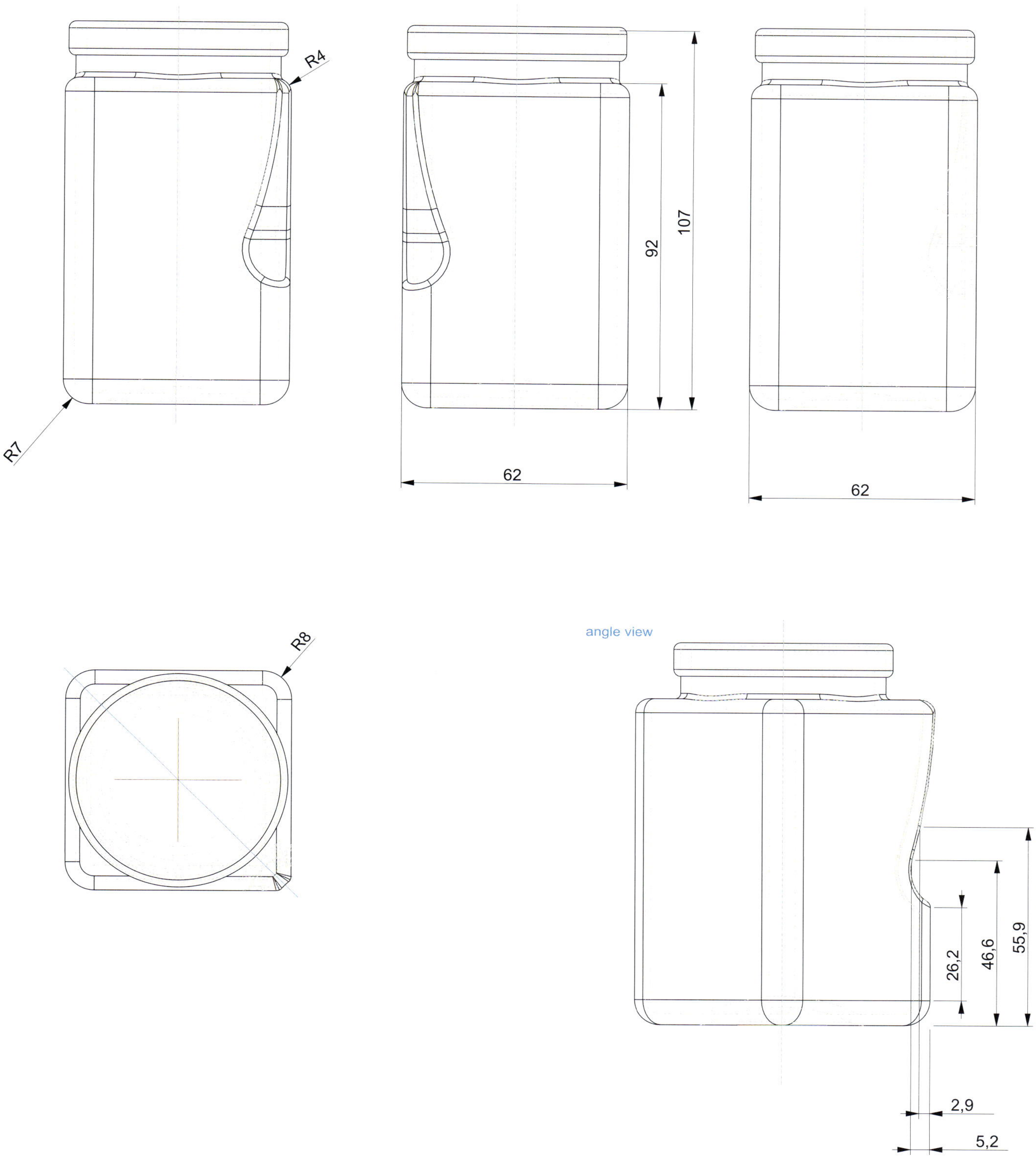

Jar 300g.
Assembled with cap
(specification number 58 size 60.15 mm)
Jar thikness 3 mm
Volume 254 (+-10) ml

JUST LAID

Thought

As part of our ongoing work with Noble Foods, we were asked to develop concepts that explored hypothetical positionings for how eggs could better satisfy or respond to consumer needs. One of the many thoughts we had was how this commodity category could respond better to concerns related to CSR issues, such as food miles and related freshness. We were intrigued by the notion that, while early adopters of CSR issues find them easy to support, it is harder to convert mass-market consumers unless there is something in it for them.

Solution

Our solution was to emphasise the benefit of locally laid eggs to consumers through a cheeky brand name that reinforces the idea of freshly laid eggs straight from the hen. The creative execution is a selection of playful caricatures of hens laying eggs with three different poses/expressions for each of the three variants.

Unique features

The Just Laid concept uses standard egg box packaging in a unique way that allows the consumers to interact with the pack and design. To further highlight the freshly laid concept, the fastening on the box becomes a representation of the egg itself, allowing consumers to connect with the natural process of the eggs 'being laid'!

Design Agency:
Springetts Brand Design Consultants

Designer:
Kevin Daly

Client:
Noble Foods

Illustrator:
Pete O'Connor

PIRATE - FOOD & DRINK

Designer:
Marko Danilović

The project consists of four different health food products under the same production company I called: Pirate – Food & Drink. The idea of this issue is to follow the life of a pirate and in order to entertain a younger audience, I aspire to revive the spirit of all those myths and legends we all loved when we were kids. Pirate – Food & Drink is a student project and it comprises three different flavors of chocolate, corn flakes, snacks and refresher – drink.

The chocolate pack has educational character and it is enriched with historical information about the pirate that inspired the product. Also there is a game included, a treasure map that leads to one special chocolate cube with a special aroma.

The corn flakes pack is defined in three flavors that are colored differently and have their own names based on the pirates that travelled through history as the real historical figures or fictive characters. Every pack is followed by illustrated skull (historically named "Jolly Roger") that every pirate ship had on its flag as a trademark. Customized by its shape and elements, it symbolized a certain guild. By cutting off a specific piece of this packing, you can get your own pirate hat with a symbol of each aroma (or "guild").

Under the name "Snacks" you will find three separate pieces, two boxes (food products in one of them) and one extra part to help them form a visually interesting shape of a "spyglass", made for user's entertainment.

Refresher is a line of three packagings of refreshing drinks. It consists of a plastic bottle and a wrapper that suggests to the salesperson to present the product in unusual way by turning the bottle upside down and hanging it on the store's shelf. By doing so, this visual form justifies the name "Peg Leg" (pirate's wooden leg).

By conjoining all the elements – treasure map, hat, spyglass and refresher the consumer is equipped with everything one pirate needs in a search for treasure and so, he is ready to choose his "guild" and join the game.

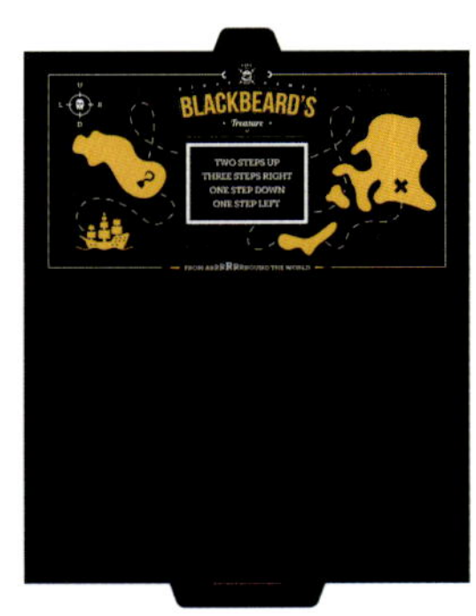

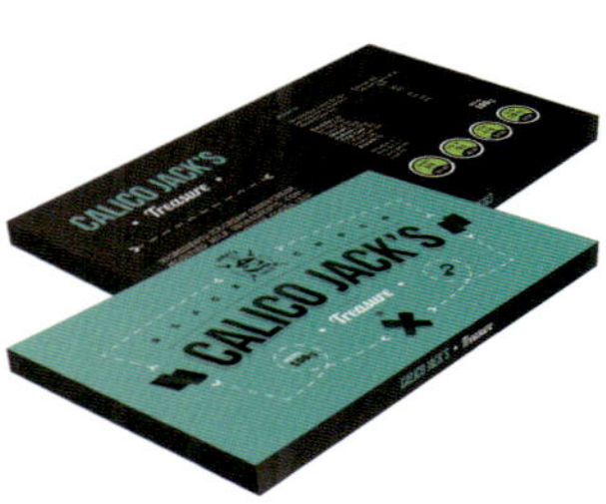

PEG LEG
Refresher
PEG LEG
Refresher
SPYGLASS
Snack's
SPYGLASS
Snack's
100g

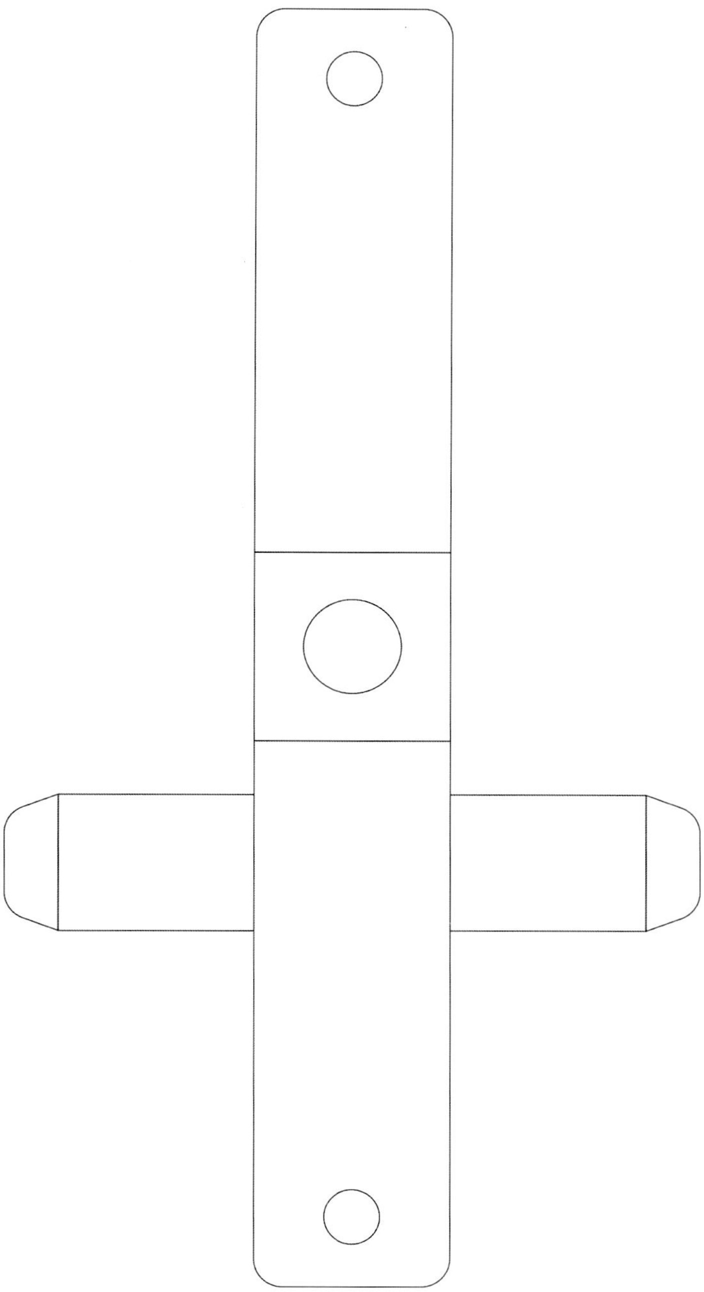

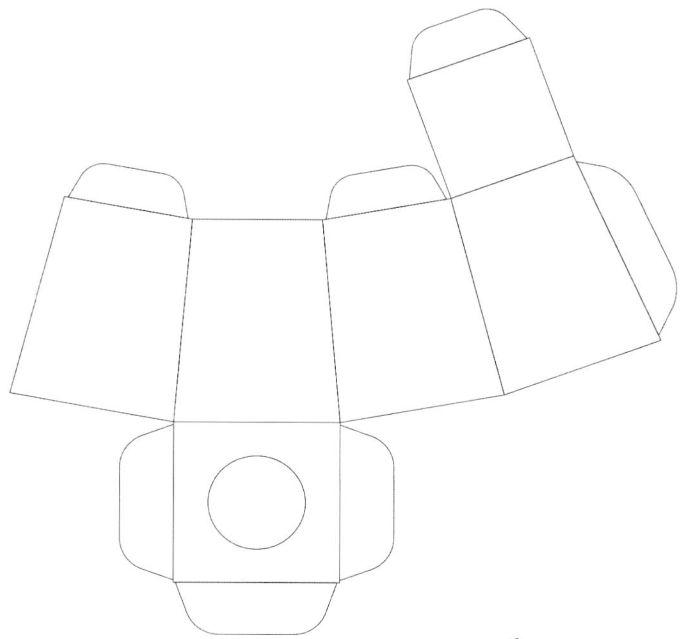

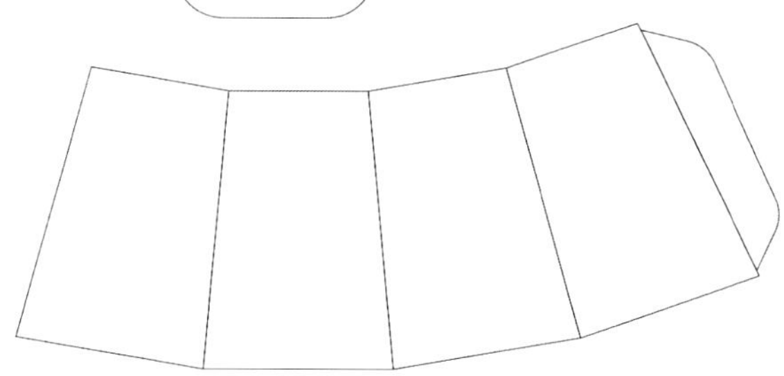

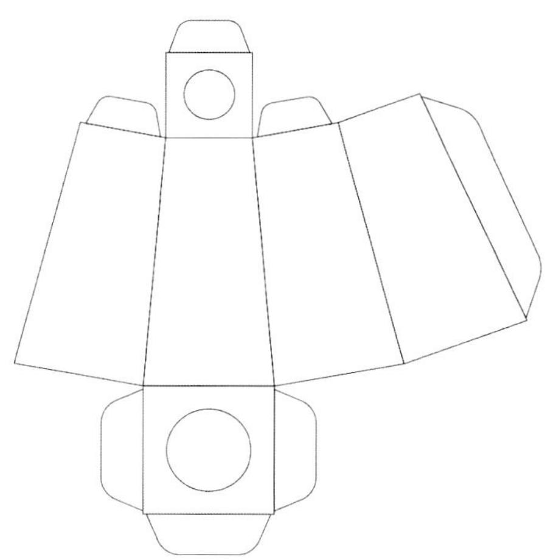

SEA
Corn Flakes
350g
SEA
Corn Flakes

SEA
Corn Flakes
SEA
Corn Flakes
350g

SEA
BEE
BUG
BUG
BUG
SEA
BEE
BEE
BEE
BUG
SEA
SEA

BUG
Corn Flakes
350g
BUG
Corn Flakes

BEE
Corn Flakes
350g
BEE
Corn Flakes

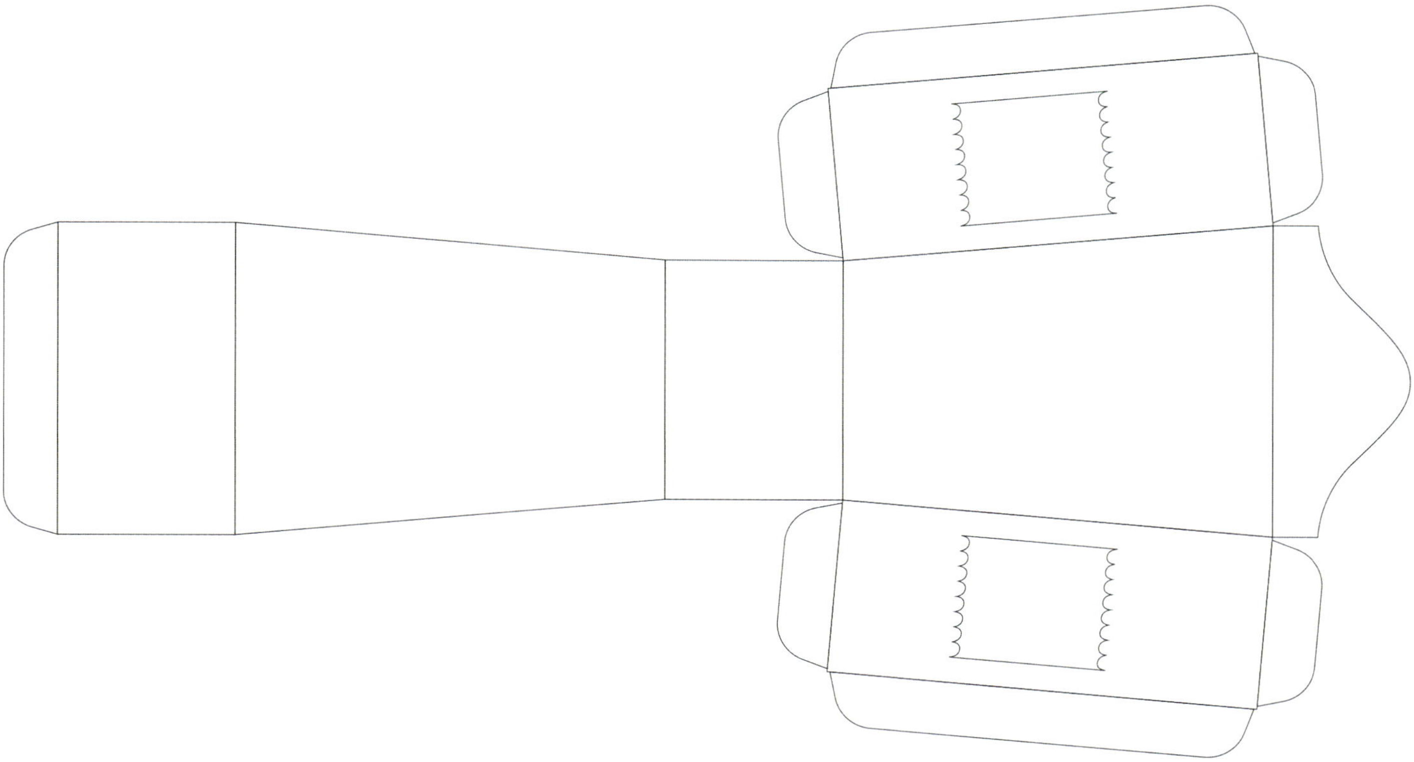

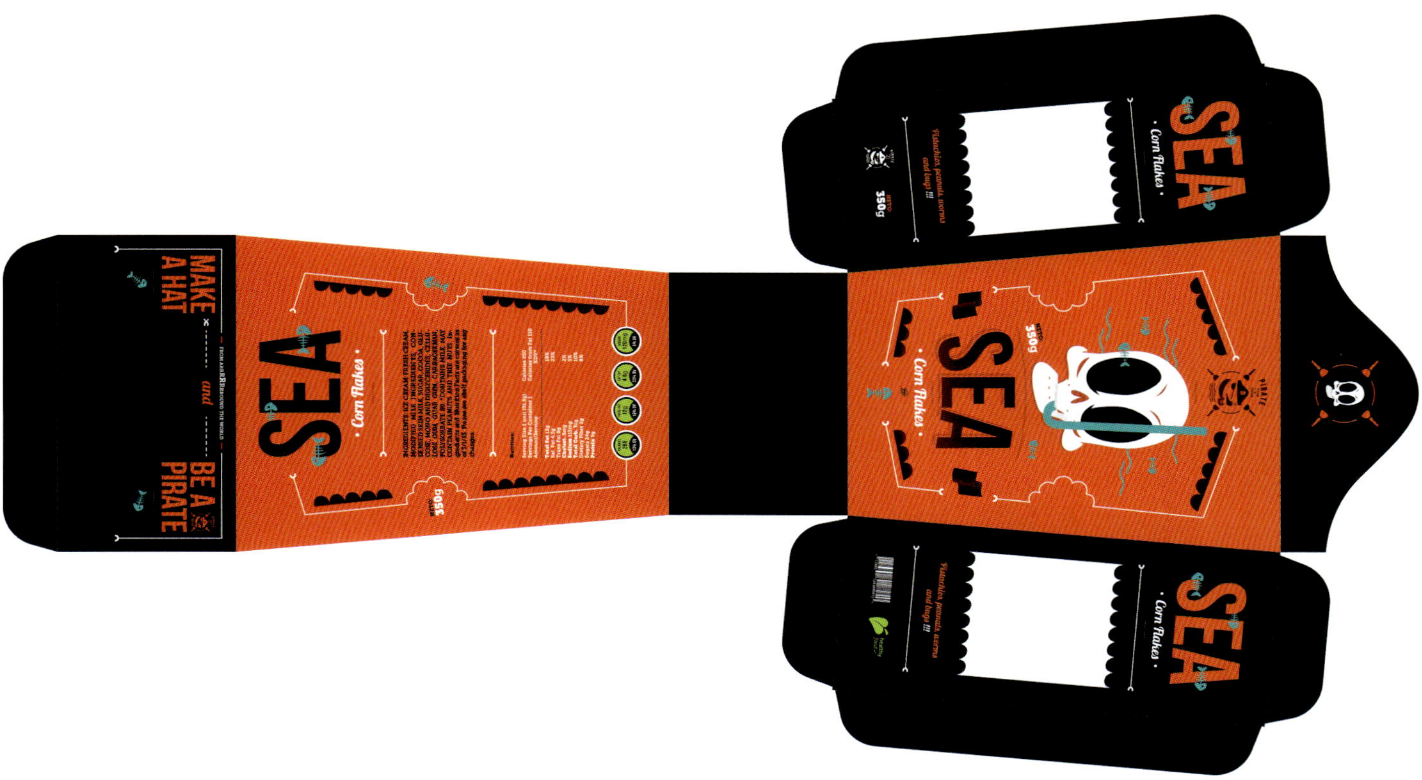
MAKE A HAT
BE A PIRATE
SEA
Corn Flakes
350g

PARTY BARON

Party Baron. Concept of the packaging for the series of fruit & berry punches.

Happy When Decrowned! Everyone who chooses alcohol drinks often imagines himself to be a connoisseur. "A hint of baked apple", "the scent of plum and smoke", "shades of rose and apricot", "velvet taste". And it is also often that only one drink is enough to take all the arrogance to a back seat — people can just enjoy life. The concept of packaging for the series of fruit & berry punches called "Party Baron" is the epitome of this phenomenon. All the lordliness disappears when the cover is open — you are decrowned and happy. You are Party Baron!

Design Agency:
Brandiziac Branding Agency

Creative Director:
Artem Shutov

Creative Idea:
Dmitriy Ardeyev, Olga Mosina, Sergey Ermakov

Designer & Illustrator:
Olga Mosina

Copywriter:
Dmitriy Ardeyev, Daria Hlyapova

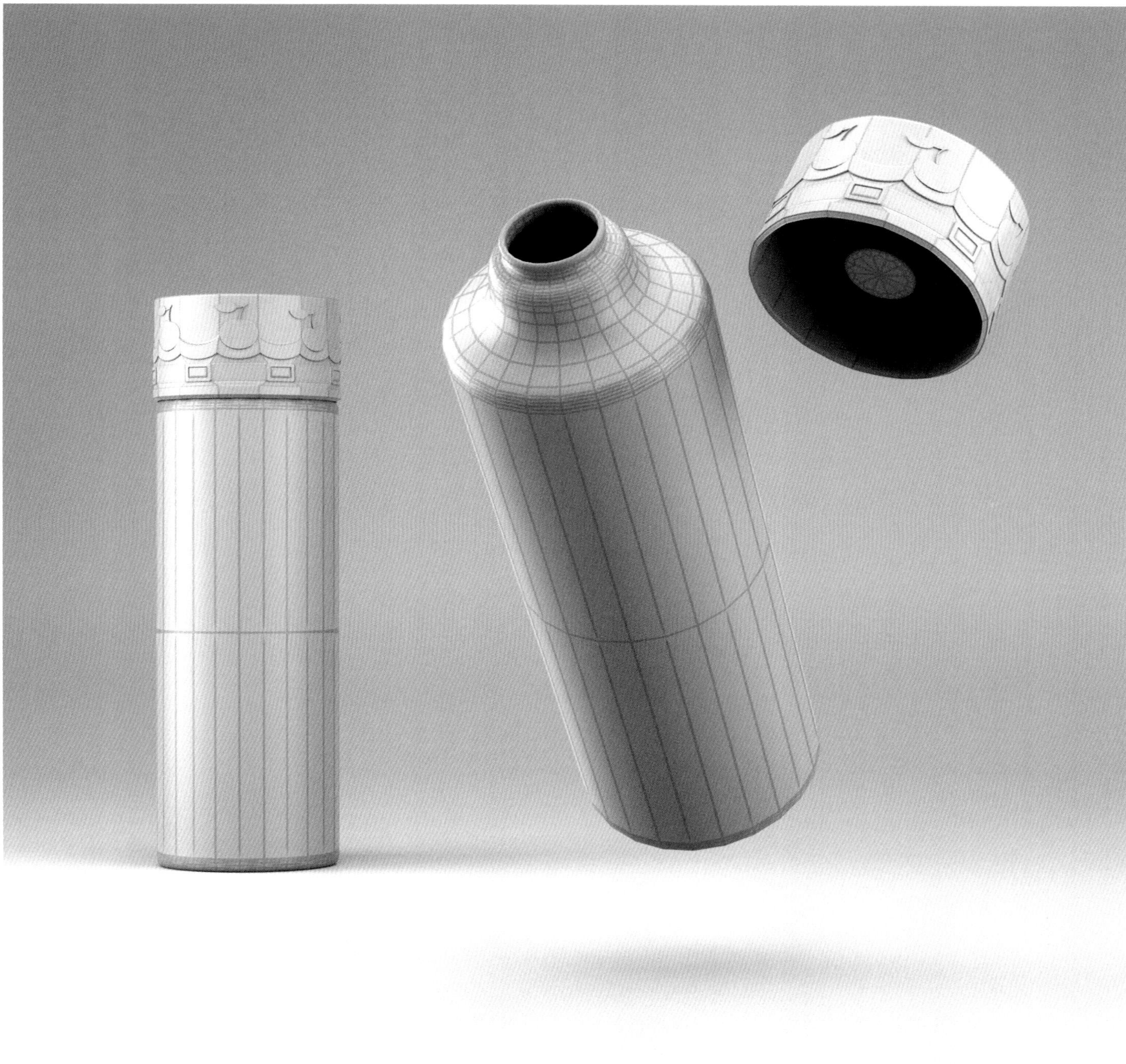

ROOTS TO GO

Roots to Go was created to bring the true flavor of the earth's treasures to people who like to eat well but value health and well-being. By transforming highly nutritive Brazilian roots into top quality snacks, the company became consolidated as a company dedicated to nature and nutrition, besides boasting a modern concept and language.

Brazilian in origin, the company sought in this country rich in foods and nature the source of inspiration and the origin of its products. Roots that are tasty, nutritious and essential for a day-to-day diet, while helping to furnish vitamin requirements and many other benefits including fiber and minerals.

Roots to go is adding these natural nutrients and flavors introducing them into hectic city life. More than just a snack, roots to go is health in a package that is arriving quickly, informally and in a relaxed manner, pleasing consumers of every age.

And crazy about chips, they went beyond to bring to everyone ... "The best roots of Brazil in just one package", and we worked to express all this modernity and values in the design of the package so that to highlight this new product in the market in Brazil and around the world.

For that we used unusual illustration in the front of the package and vibrant colors to remind the happiness and cheerful brazilian way of life. and we worked to express all this modernity and values in the design of the package so that to highlight this new product in the market in Brazil and around the world.

Design Agency:
Spice Design

roots
vegetable chips
to
go
• AMBER SWEET POTATO
• YUCA • PURPLE SWEET POTATO
• RUBY YUCA • BRAZILIAN TARO
CARÁ, BATATA-DOCE E MANDIOCA FRITOS EM FATIAS
NET WT 2oz / PESO LÍQUIDO 56g
roots
togo
YUCA
MANDIOCA
BRAZILIAN TARO
CARÁ
AMBER SWEET POTATO
RUBY YUCA
www.rootstogo.com.br

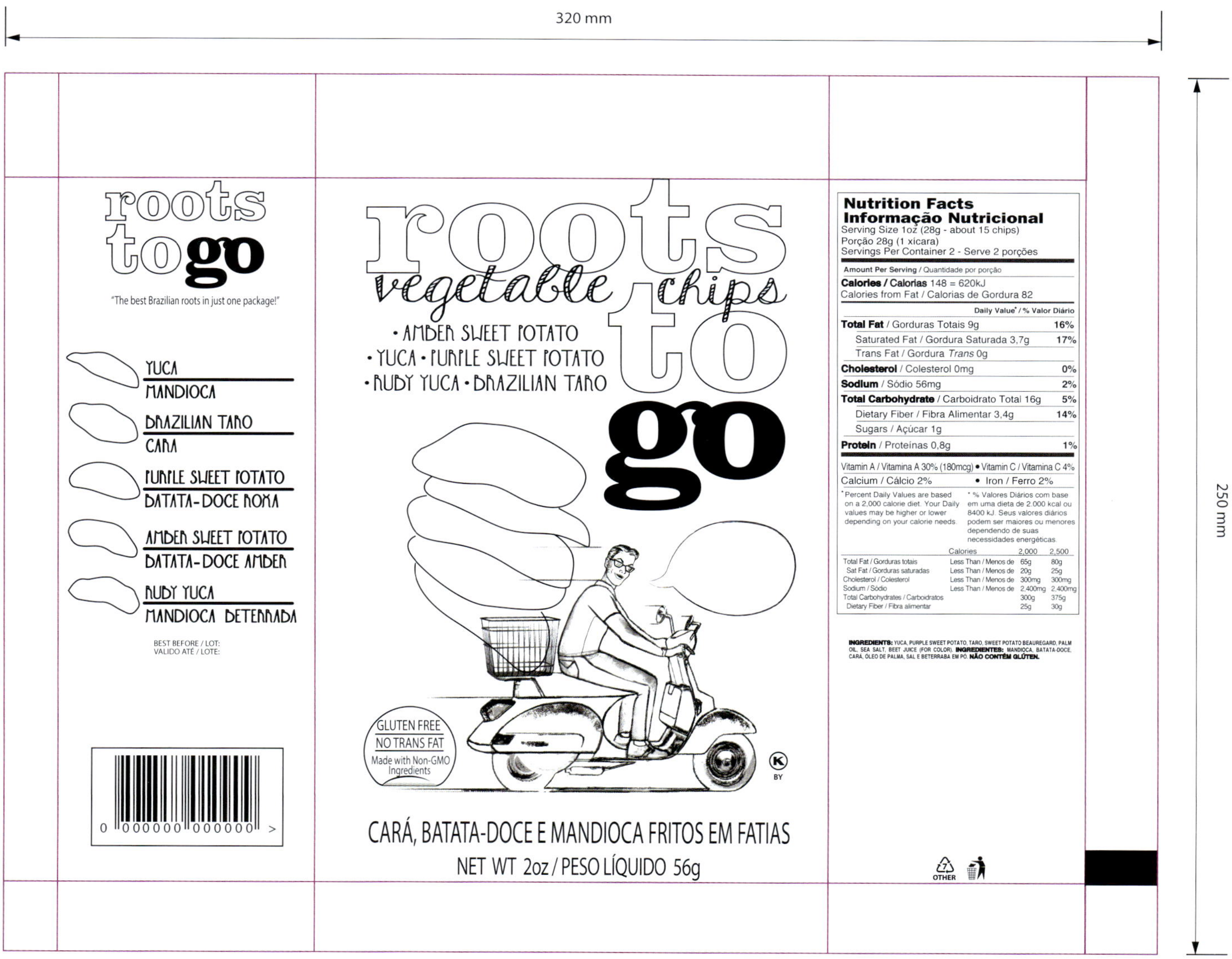

320 mm
250 mm
roots
togo
"The best Brazilian roots in just one package!"
YUCA
MANDIOCA
BRAZILIAN TARO
CARÁ
PURPLE SWEET POTATO
BATATA-DOCE ROXA
AMBER SWEET POTATO
BATATA-DOCE AMBER
RUBY YUCA
MANDIOCA BETERRABA
BEST BEFORE / LOT:
VALIDO ATÉ / LOTE:
roots
vegetable chips
to
go
• AMBER SWEET POTATO
• YUCA • PURPLE SWEET POTATO
• RUBY YUCA • BRAZILIAN TARO
GLUTEN FREE
NO TRANS FAT
Made with Non-GMO Ingredients
CARÁ, BATATA-DOCE E MANDIOCA FRITOS EM FATIAS
NET WT 2oz / PESO LÍQUIDO 56g
Nutrition Facts
Informação Nutricional
Serving Size 1oz (28g - about 15 chips)
Porção 28g (1 xícara)
Servings Per Container 2 - Serve 2 porções
Amount Per Serving / Quantidade por porção
Calories / Calorias 148 = 620kJ
Calories from Fat / Calorias de Gordura 82
Daily Value* / % Valor Diário
Total Fat / Gorduras Totais 9g 16%
Saturated Fat / Gordura Saturada 3,7g 17%
Trans Fat / Gordura Trans 0g
Cholesterol / Colesterol 0mg 0%
Sodium / Sódio 56mg 2%
Total Carbohydrate / Carboidrato Total 16g 5%
Dietary Fiber / Fibra Alimentar 3,4g 14%
Sugars / Açúcar 1g
Protein / Proteínas 0,8g 1%
Vitamin A / Vitamina A 30% (180mcg) • Vitamin C / Vitamina C 4%
Calcium / Cálcio 2% • Iron / Ferro 2%
* Percent Daily Values are based on a 2,000 calorie diet. Your Daily values may be higher or lower depending on your calorie needs.
* % Valores Diários com base em uma dieta de 2.000 kcal ou 8400 kJ. Seus valores diários podem ser maiores ou menores dependendo de suas necessidades energéticas.
Calories 2,000 2,500
Total Fat / Gorduras totais Less Than / Menos de 65g 80g
Sat Fat / Gorduras saturadas Less Than / Menos de 20g 25g
Cholesterol / Colesterol Less Than / Menos de 300mg 300mg
Sodium / Sódio Less Than / Menos de 2,400mg 2,400mg
Total Carbohydrates / Carboidratos 300g 375g
Dietary Fiber / Fibra alimentar 25g 30g
INGREDIENTS: YUCA, PURPLE SWEET POTATO, TARO, SWEET POTATO BEAUREGARD, PALM OIL, SEA SALT, BEET JUICE (FOR COLOR). INGREDIENTES: MANDIOCA, BATATA-DOCE, CARÁ, ÓLEO DE PALMA, SAL E BETERRABA EM PÓ. NÃO CONTÉM GLÚTEN.
OTHER

ECOLEAN OLFRUTE

We conceptualized, art directed & designed a package artwork for Engro Foods, a consumer foods company based in Pakistan. The brief was to develop a refreshing new look for one of their existing brands – Olfrute – that would stand out but also showcase their new, transparent package Ecolean Air Clear.

Design Agency:
Binalogue

Art Director:
David Carrizales, Marcus Stenbeck

Designer:
Jorge Artola, Inzar Zamora

Production Company:
Ecolean

Client:
engro foods

Ecolean® Air Aseptic 200 Front
Ver. nr: 4100804-F 2011-11-18
No printing area
INSERT STRAW FOR THE FUN TO BEGIN!
OLFRUTE
LEMONADE
200 ml
NO PRESERVATIVES

Ecolean® Air Aseptic 200 Back
Ver. nr: 4100804-F 2011-11-18
No printing area
INSERT STRAW FOR THE FUN TO BEGIN!
Optional, area for straw application
REMOVE THE STRAW
click!
TEAR THE CORNER
INSERT THE STRAW
GRIP THE PACK BY AIRHANDLE
OLFRUTE
Olfrute tastes, smells and even sounds like what it stands for! All fruit, no preservatives. With the new see-through pack you can even see it! Just put the straw in the spout and enjoy. Let it refresh you over!
engro foods

ecolean
a lighter approach to packaging
ecolean
Non printing area
Ecolean Air Aseptic 200/Bottom
Rev. nr 4100992-A_Bottom/2011-11-25

HEINZ

The design highlights Heinz's use of natural ingredients and their particularly careful attitude towards their ketchup. Kraft paper was the ideal material to convey these features visually. The Victorian printing-press gave the design a sense of heritage. Heinz tomato ketchup is now on the shelves throughout Russia.

Design Agency:
Depot WPF

Designer / Art Director:
Maria Ponomareva

Client:
Heinz

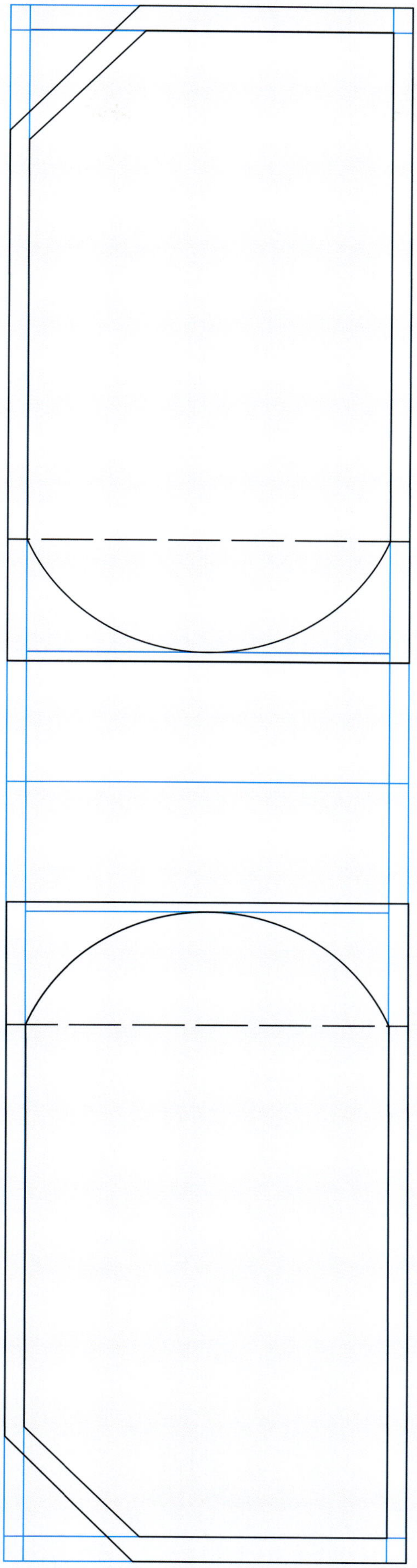

Width of laminated film — 498 mm.
Step of print — 125 mm.

OURSON LUNE DE MIEL

The Lune de Miel® launched a new product aimed at families: "the Teddy Bear". The agency designed the volume by imagining a mascot-packaging, to be treated as a character. A muzzle, a forehead, eyebrows, ears, a smile... each element was designed as a volume and then reinforced with screen printing to give the Teddy Bear a unique personality and look.

This mascot-packaging is large, round and endearing. It also meets all the industrial technical requirements. Its stability was optimised for production lines, palletization and shelf placement. The volume design includes a standard cap, already used on an economical-size squeezer, that allows for economies of scale. And what's more, the Teddy Bear shape and material were chosen so that the packaging could have better "squeezability". The agency guided the brand Lune de Miel® by collaborating on industrial development and integrating this creation into manufacturing lines.

Design Agency:
LOGIC DESIGN

Designer:
Jérôme Lanoy

Client:
Lune de Miel

Coupe A-A

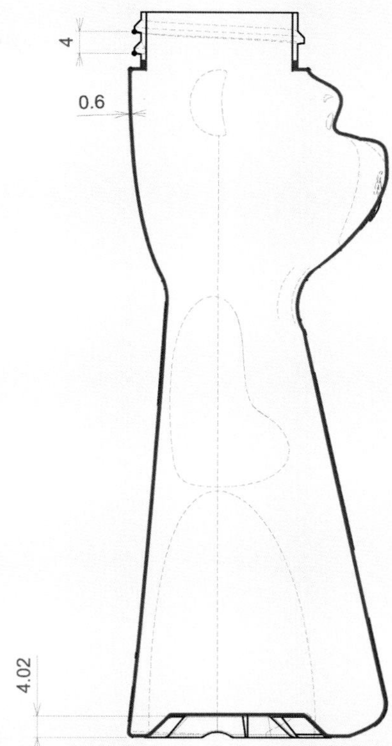

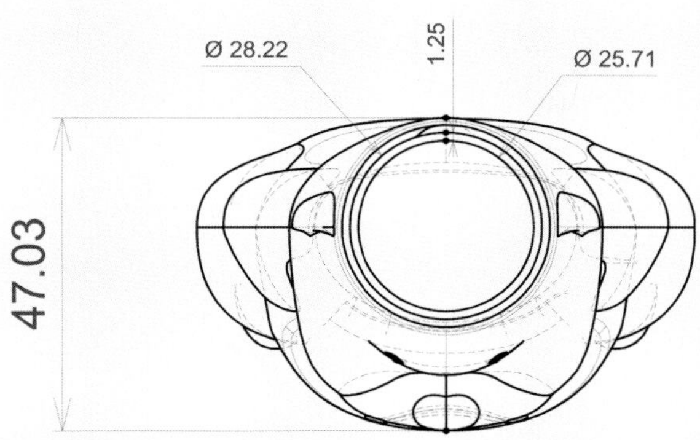

BEVERAGE PACKAGE

RICARD DUO

Richard has recently commissioned architecture and design duo Jakob + MacFarlane to create a new carafe for the brand. The pair chose to develop two carafes, to symbolize the strong union of Ricard and water – the two indissociable components of the famous drink.

Called Duo and carved in blocks of glass, the clear water bottle is made to look like a piece of cool ice, while the bright yellow one protects the precious liquid.

Design Agency:
Jakob + MacFarlane

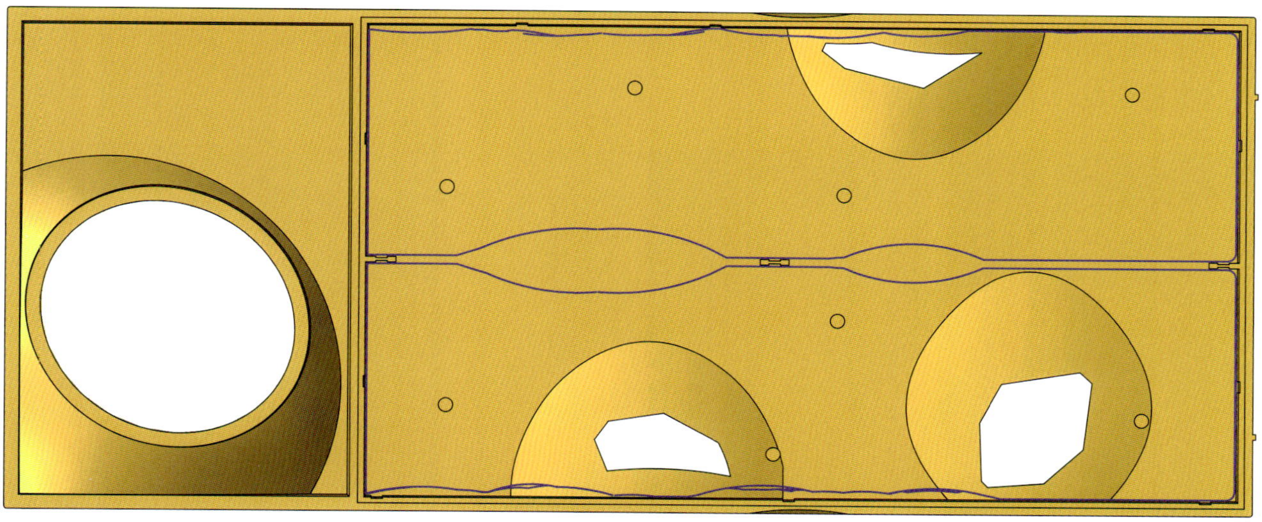

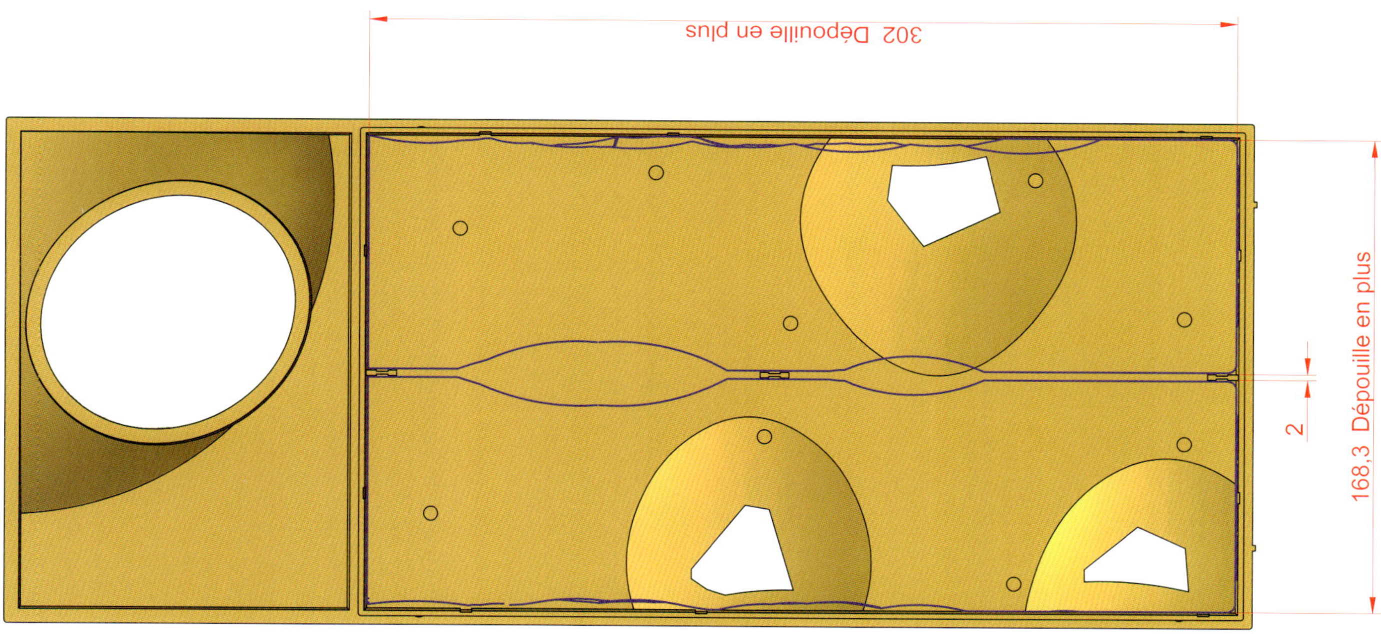
302 Dépouille en plus
168,3 Dépouille en plus
2

RICARD

RICARD

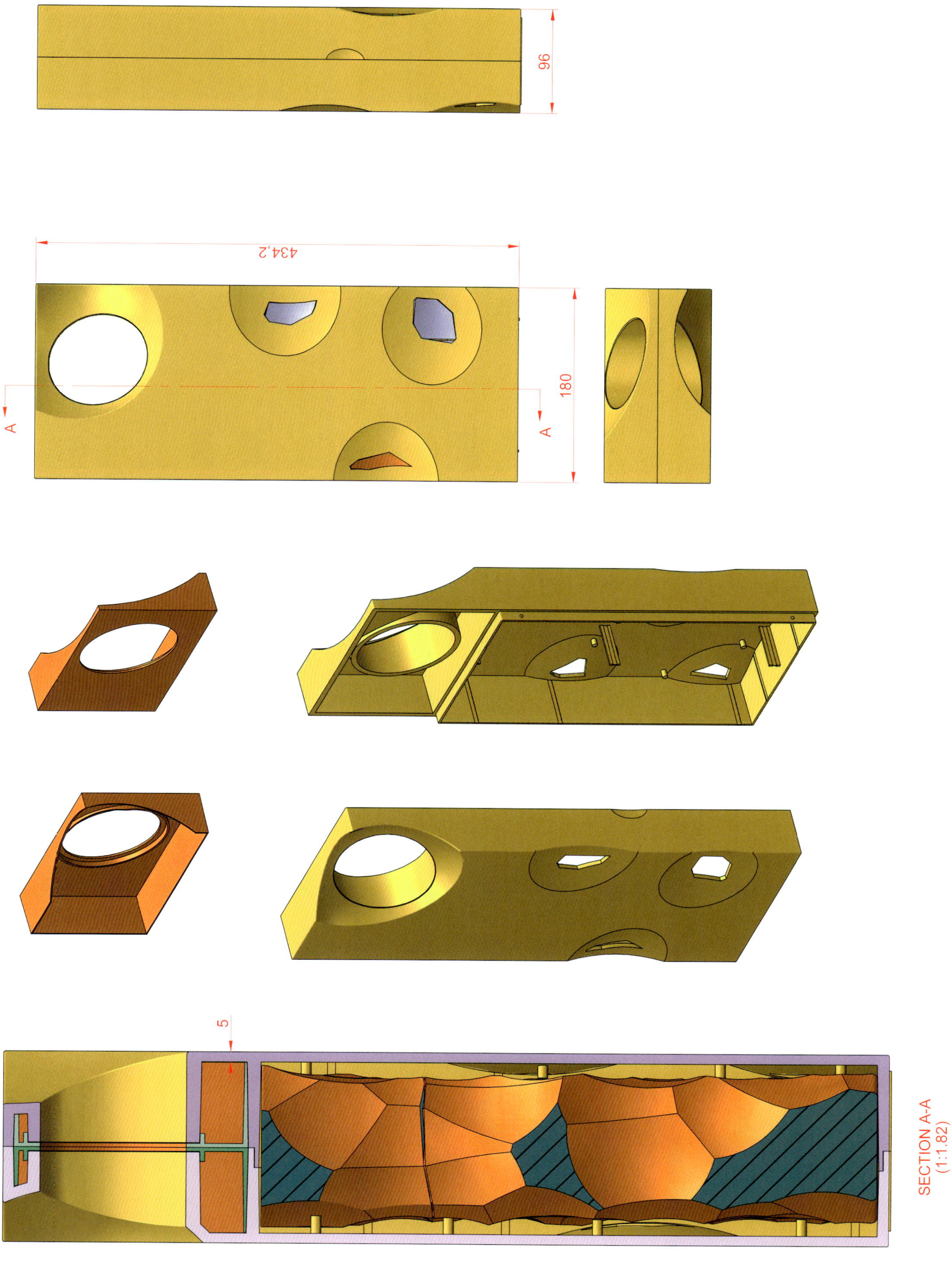
96
434,2
180
A
A
5
SECTION A-A
(1:1.82)

WILSON AND SCHWARTZ

These Northern Ohio master distillers honed their craft over the years with family recipes. In 2005, they opened their own microdistillery. To reflect the simple methods and pure ingredients used in their distilling process, we used a one-color system of screen printed bottles and letterpress collateral. Since they were on a limited budget, we chose off-the-shelf bottles and put our money into the screen printing which really elevated the packaging.

Design Agency:
C.Witzel

Designer:
Chad Witzel

Client:
Wilson and Schwartz

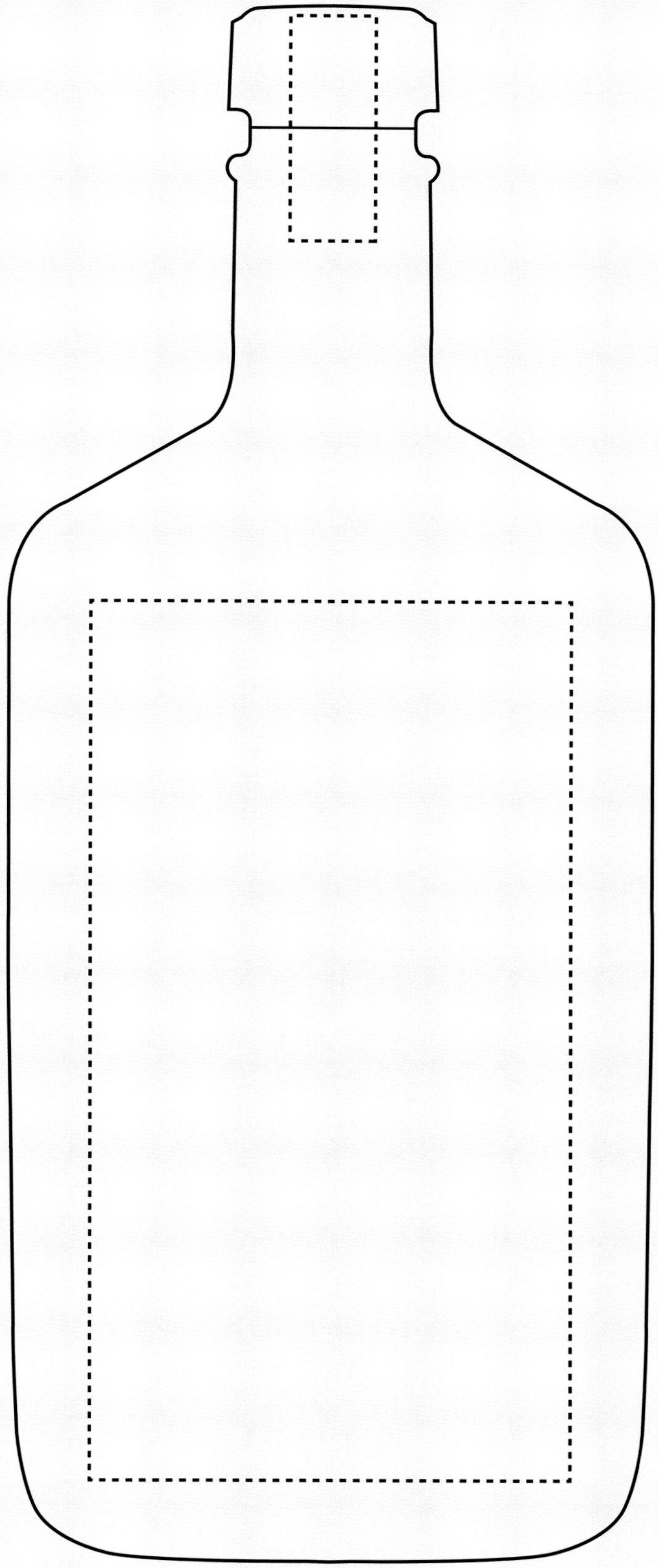

ANESTASIA VODKA BRANDING & PACKAGING

AnestasiA Vodka has opened up a new chapter in the world of alcoholic spirits. A spirit this innovative requires an equally visionary bottle. Unique and crystalline, the faceted surface gives it an ultra-futuristic feel. The hard-edge, asymmetrical design also provides an ease of use and intuitive functionality. The beauty of the bottle makes it a perfect table top piece, accenting the environment with a look embodying cutting-edge style. These factors contribute to AnestasiA Vodka being a new sensory experience unlike any other alcoholic product.

Design Agency:
Karim Rashid

Designer:
Karim Rashid

Client:
AnestasiA Vodka

Photography:
Karim Rashid

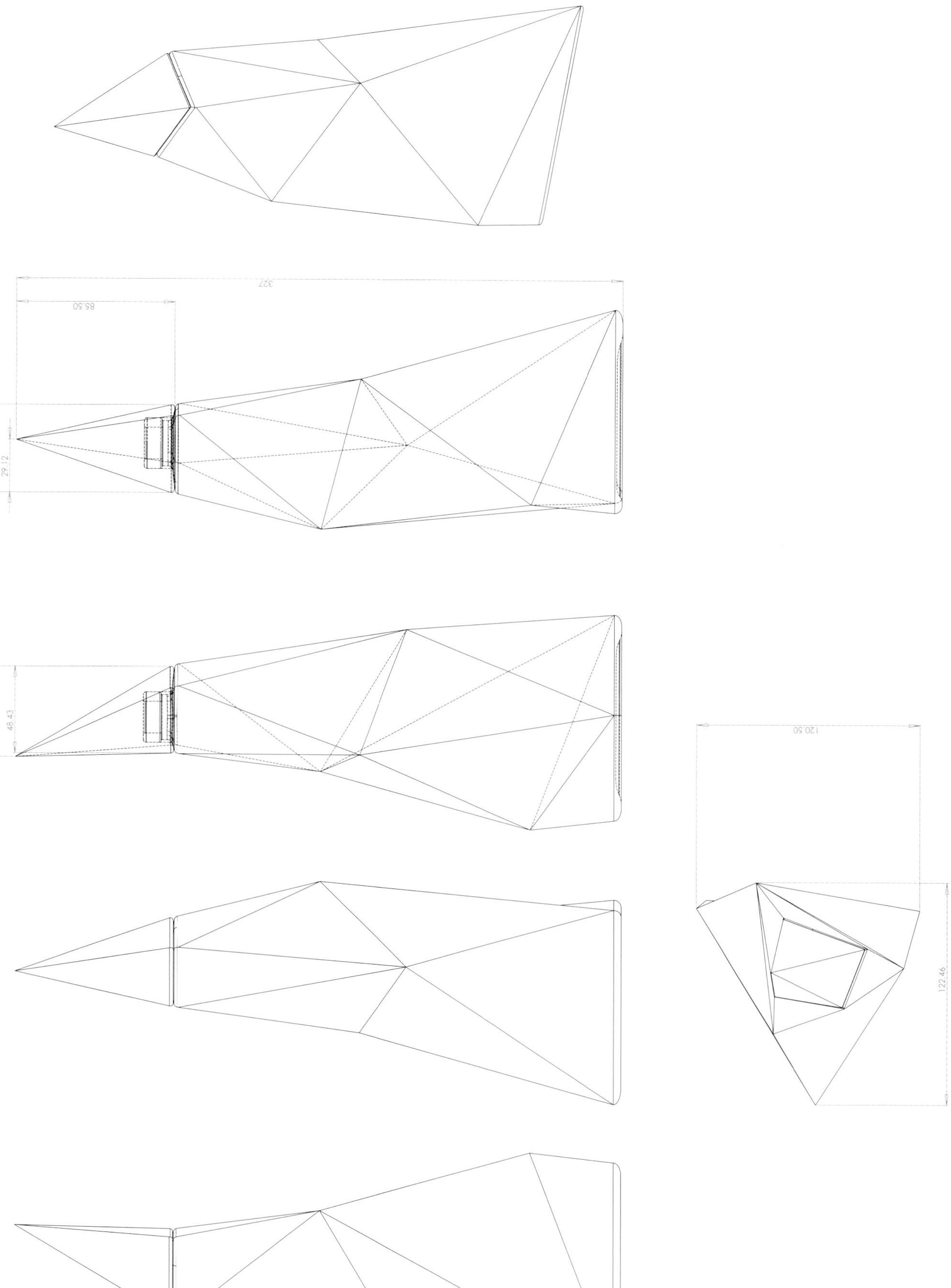
327
85.50
48.40
29.12
50.54
48.43
120.50
122.46

Designer:
Fabian Fohrer

MITTELMASS

Neither the most traditional nor the one and only. Not the greenest, not the most golden. No king and certainly not a pearl of nature[1]. Mittelmass does not need superlatives. It's a beer as any other. Water, malt and hops, nothing else.

Does advertising influence the character of a beer? Probably not. But it affects your imagination. So shouldn't beer preferably be liked by its taste, instead by its description?

To not manipulate this taste, Mittelmass (English: »average«) defines itself with understatements and allows space for a personal and individual impression.

[1] German beer advertising slogans

[2] ingredients of beer, as it's defined in the German purity law

ø
Gebraut nach dem deutschen Reinheitsgebot.
Zutaten: Wasser, Malz, Hopfen
Weder das traditionellste noch das einzig Wahre. Nicht das Grünste, nicht das Goldenste. Kein König und bestimmt auch nicht die Perle der Natur. Mittelmaß braucht keine Superlativen. Es ist ein Bier wie jedes andere. Wasser, Malz und Hopfen. Mehr nicht.
Mittelmaß Biermanufaktur
— fabianfohrer.com/mittelmass
Mindestens haltbar bis:
siehe Flaschen
3 × ℮ 0,33l
alc. 4,8% vol
★★★☆☆
Wasser, Malz, Hopfen.
Mehr nicht.
Bier seit Zweitausendfünfzehn.
mittel
maß
drei mal
null komma drei drei liter

COSMOS

This is the wine bottle design, motion design packaging for Cosmos.

Designer:
Ion Lucin

SPECIAL BOTTLE DESIGN
E-PAPER FLEXIBLE DISPLAY
CIRCUIT BOARD
BATTERY
BACKLIT TEXT
SPECIAL BOTTLE DESIGN
E-PAPER FLEXIBLE DISPLAY
CIRCUIT BOARD
COSMOS
BATTERY
BACKLIT TEXT

TELAVERA TEQUILAS

I was very inspired by the culture, arts and history of Mexico, specifically Talavera. Talavera is the word used to identify the beautiful handmade pottery and fine ceramics that are trademarks of Mexican craftsmanship. Just as Mexican Talavera was created with strict standards and regulations in order to preserve the expert nature of its craft, I expect every bottle of my tequila to be crafted in this same way; with precise ingredients and regulations that go into making every bottle a beyond tasteful experience. To give each bottle of tequila the utmost feeling of prestige, I incorporated an important element in Mexican culture and history: Mayan gods. Religion played an enormous role in talavera ceramics so I wanted to make these characters come to life on each bottle in a sophisticated, yet lively way. The gods that are seen on the substrates are: Yum Kaax, The God of Nature; Chaac: The God of Rain; and Kinich Ahau: The God of the Sun.

Design Agency/Designer:
Hannah Hart Design / Hannah Hart

Photography:
Hannah Hart

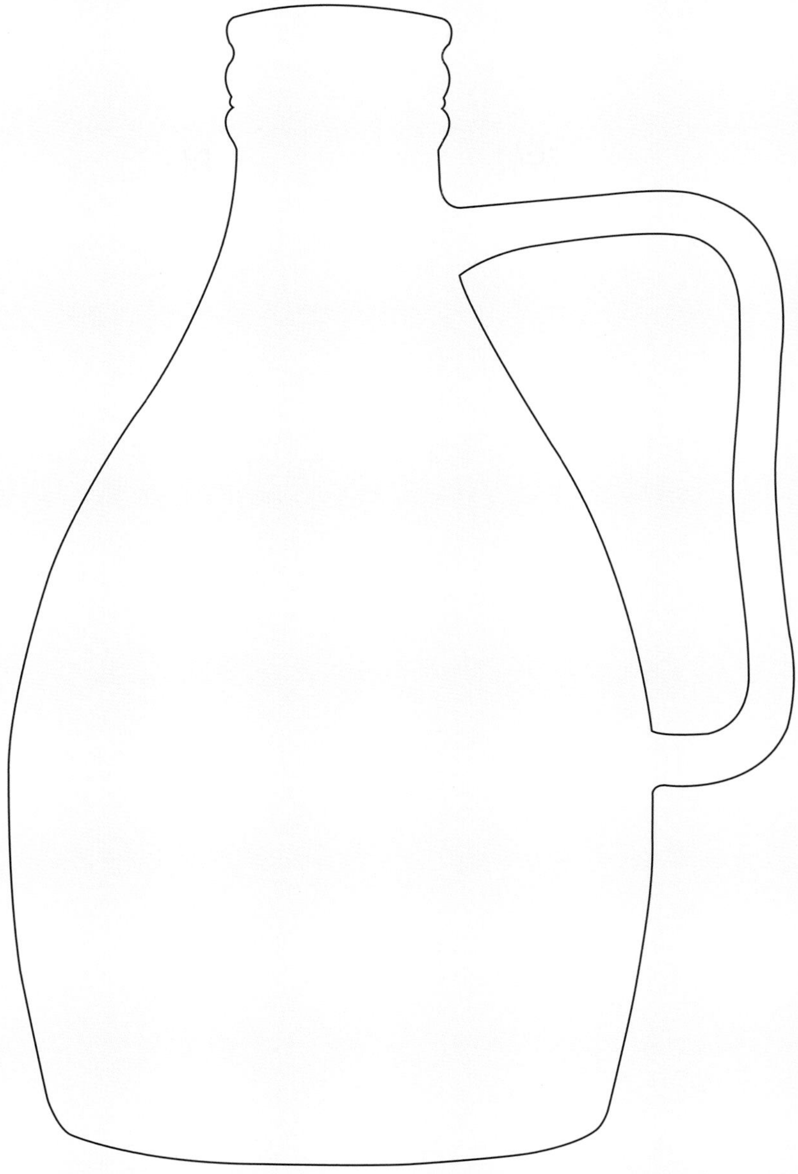

telavera
TEQUILAS
TEQUILA AÑEJO
chaac
MADE IN MEXICO
40% ALC BY VOL. | 1 LITER

telavera
TEQUILAS
TEQUILA SILVER
MADE IN MEXICO
40% ALC BY VOL. | 1 LITER

telavera
TEQUILAS
TEQUILA REPOSADO
yum kaax
MADE IN MEXICO
40% ALC BY VOL. | 1 LITER

CUBANISTO

Discover the spirit within. Cubanisto is a rum flavoured premium beer with a fresh taste of citrus, orange zest, lime and an aroma of caramelised cane sugar and treacle. Appealing rum flavoured beer, with a surprising benefit!!

Design Agency:
Mountain Design

Designer:
Jon Sonneveld

Account:
Jorg Weppelink

Client:
AB Inbev

Illustrator:
Jon Sonneveld/Kate Varaday

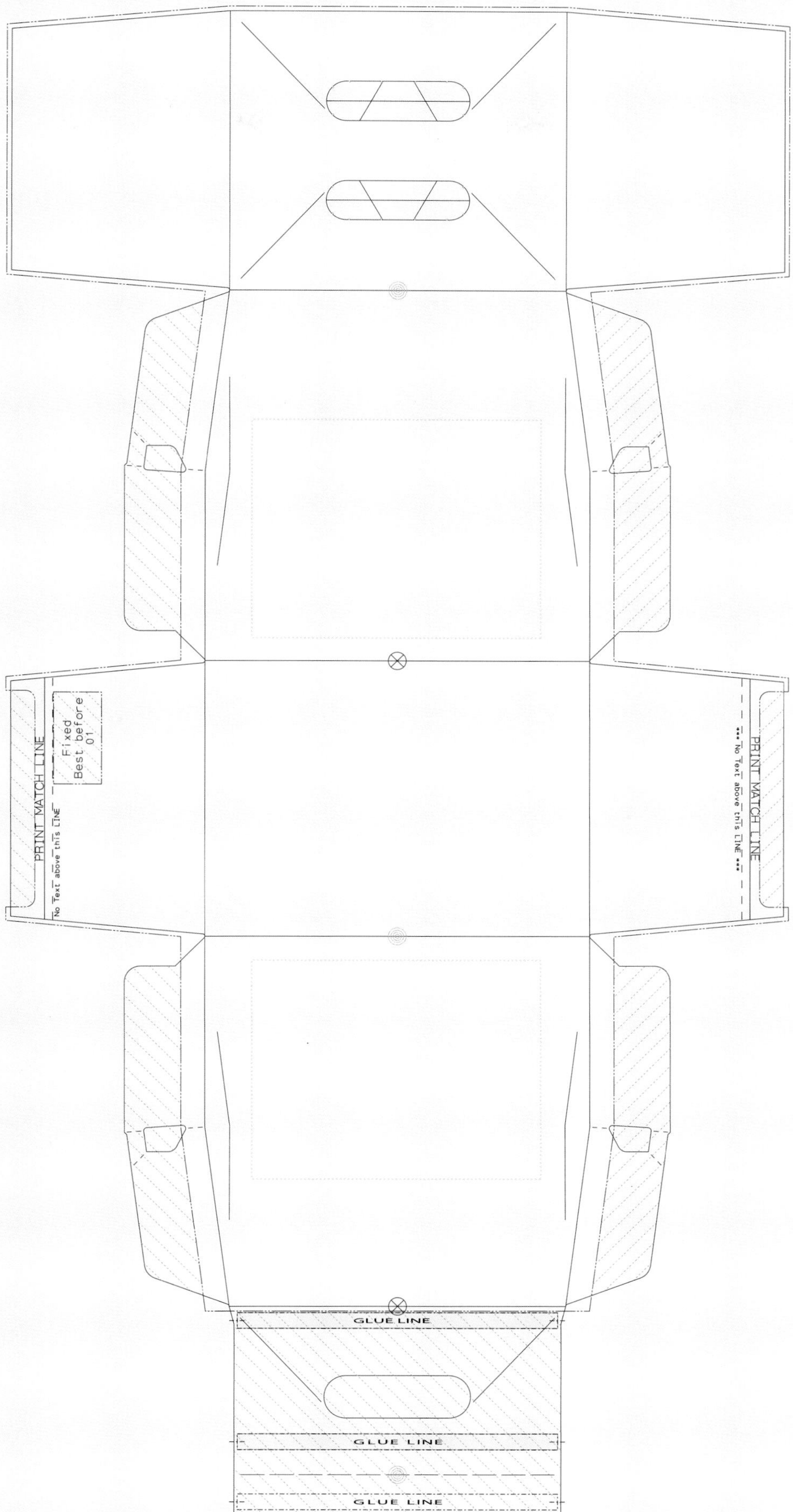

Fixed
Best before
01
PRINT MATCH LINE
No Text above this LINE
PRINT MATCH LINE
*** No Text above this LINE ***
GLUE LINE
GLUE LINE
GLUE LINE

CUBANISTO
El Espíritu del Caribe

CUBANISTO
RUM
FLAVOURED BEER
33
CL
Alc.
5,9%
vol.

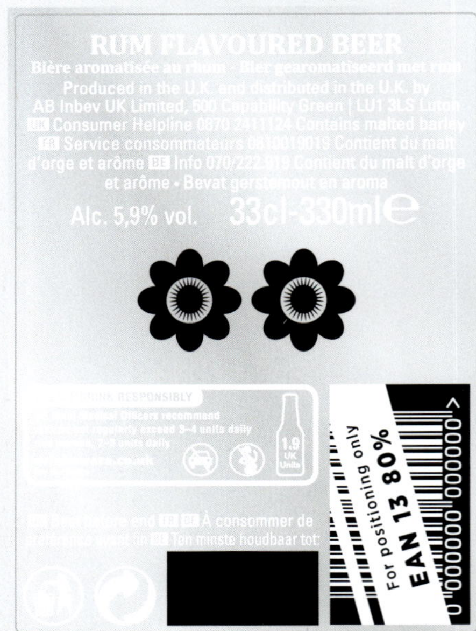
RUM FLAVOURED BEER
Produced in the U.K. and distributed in the U.K. by
AB Inbev UK Limited, 500 Capability Green | LU1 3LS Luton
Consumer Helpline
Service consommateurs
et arôme • Bevat gerstemout en aroma
Alc. 5,9% vol.
33cl-330ml℮
DRINK RESPONSIBLY
1.9 UK Units
À consommer de
Ten minste houdbaar tot:
For positioning only
EAN 13 80%

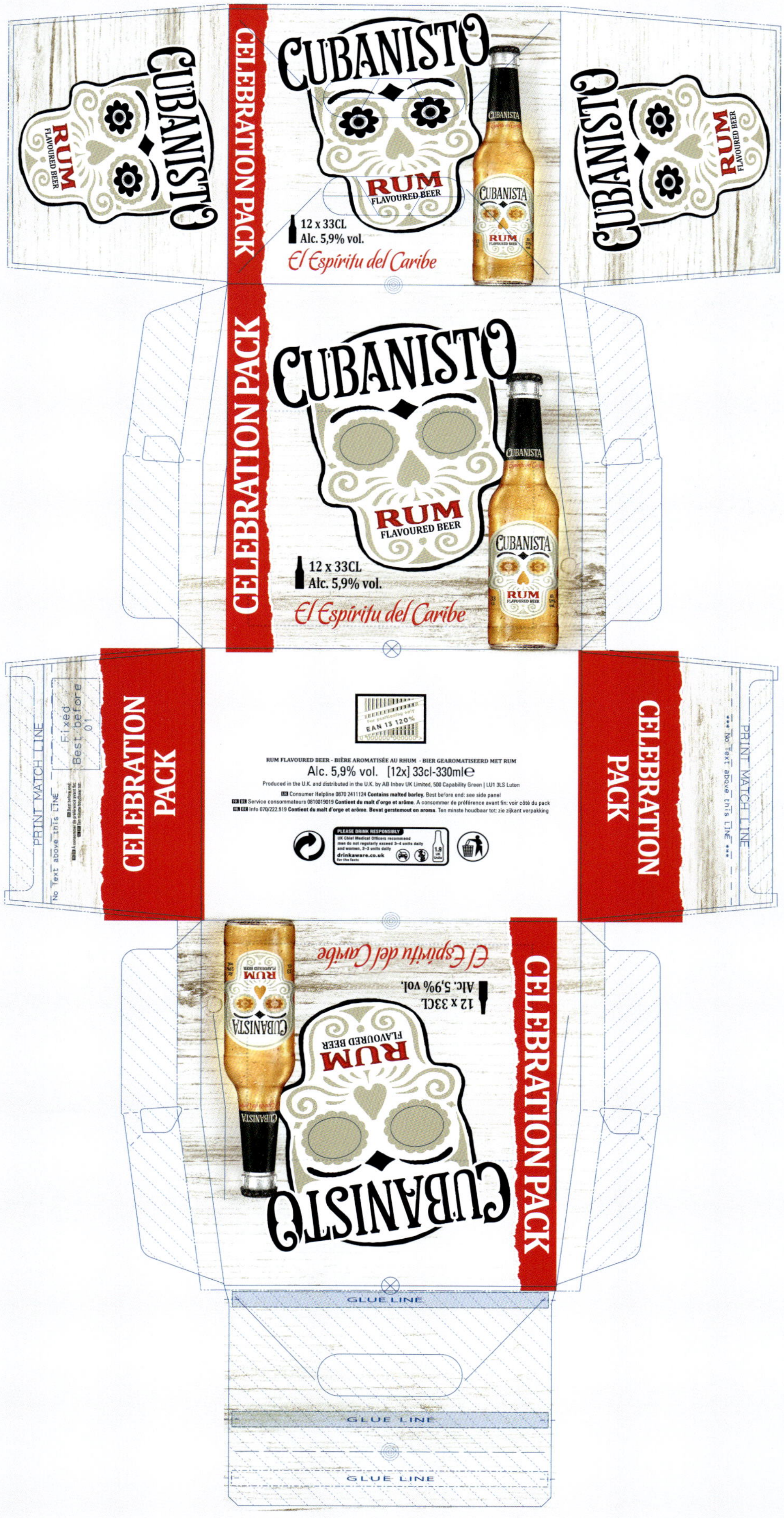
CUBANISTO
RUM
FLAVOURED BEER
CELEBRATION PACK
12 x 33CL
Alc. 5,9% vol.
El Espíritu del Caribe
CUBANISTA
EAN 13 120%
RUM FLAVOURED BEER - BIÈRE AROMATISÉE AU RHUM - BIER GEAROMATISEERD MET RUM
Alc. 5,9% vol. [12x] 33cl-330ml℮
Produced in the U.K. and distributed in the U.K. by AB Inbev UK Limited, 500 Capability Green | LU1 3LS Luton
UK Consumer Helpline 0870 2411124 Contains malted barley. Best before end: see side panel
FR BE Service consommateurs 0810019019 Contient du malt d'orge et arôme. A consommer de préférence avant fin: voir côté du pack
NL BE Info 070/222.919 Contient du malt d'orge et arôme. Bevat gerstemout en aroma. Ten minste houdbaar tot: zie zijkant verpakking
PLEASE DRINK RESPONSIBLY
UK Chief Medical Officers recommend men do not regularly exceed 3–4 units daily and women, 2–3 units daily
drinkaware.co.uk
for the facts
PRINT MATCH LINE
No Text above this Line
GLUE LINE

NALEWKA

Nalewka is a traditional Polish category of alcoholic beverage which is made by maceration of fruits, roots, spices and herbs in alcohol. In these days also industrially produced, however, the best taste comes from homemade method which is appreciated by many connoisseurs of this liquor. Our Nalewka was intended to be a gift for our friends and customers. Prepared almost entirely by hand from the beverage, to the final details of packaging.

Design Agency:
Foxtrot Studio

Designer:
Adrian Chytry, Izabella Jankowska

Client:
self-promotion

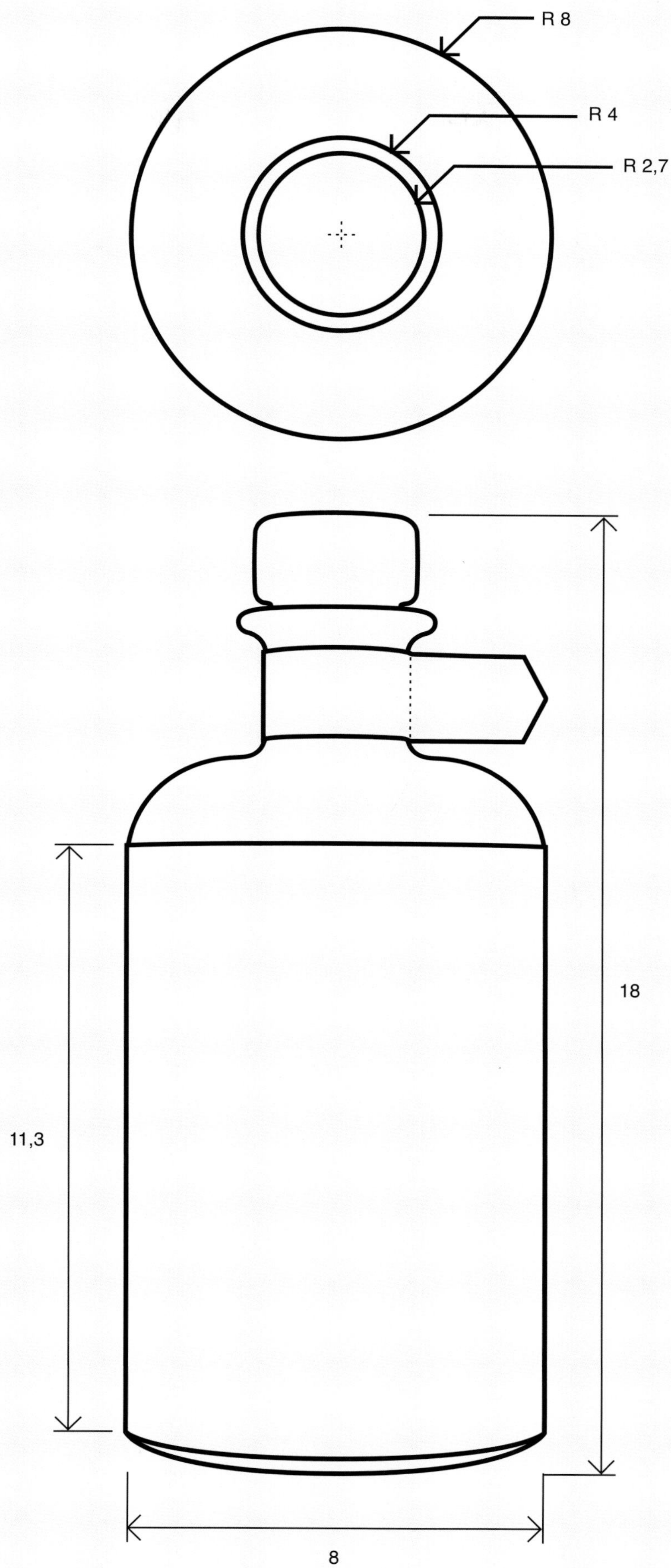
R 8
R 4
R 2,7
18
11,3
8

MAŁA PARTIA & PRYWATNY ZAPAS
FOXTROT
STUDIO
Nalewka
500ML
45% ALC./VOL.

MAŁA PARTIA & PRYWATNY ZAPAS
FOXTROT
STUDIO
Nalewka
MALINOWA
500ML
45% ALC./VOL.

Nalewka
MAŁA PARTIA
PRYWATNY ZAPAS
Nalewka

MAŁA PARTIA & PRYWATNY ZAPAS
FOXTROT
STUDIO
Nalewka
500ML
45% ALC./VOL.

Nalewka

MINIMUM

Nalewka

EL MARIACHI RED WINE COLLECTION

IKON BC was given the task of designing the labels for El Mariachi red wine series from the Sonora Province and called on the talents of Illustrator /designer Steve Simpson to work on the project alongside them.

It was important to recognise the importance of creating something that really stood out from the existing Mexican wine labels exported to the Californian and New England regions of the United States.

The El Mariachi collection is primarily targeted towards young adults, this demographic is becoming of greater importance in the red wine market. The team wanted to go beyond the conventional labels and create something that has a visual aesthetic of the brand's emotion and personality. Thus, we decided to use an accompanying neck-hanger element in the form of 3 Mariachi heads, one for each label. The three Mariachi brothers: Roberto, Chucho and Pepe, will spice up every soiree and party in a Mexican way.

0123456789012

7 5 0 2 9 1 5 8 7 0 7 1 4

Design Agency:
IKON BC.

Creative and Art Director:
Eugene Kuprienko

Designer:
Steve Simpson, Eugene Kuprienko, Anatoly Ivanov

Illustrator:
Steve Simpson

Project Director:
Dave Nichols

Papercraft Keylines:
Eugene Kuprienko

Photography:
Vasily Fetisov (FAME Studio)

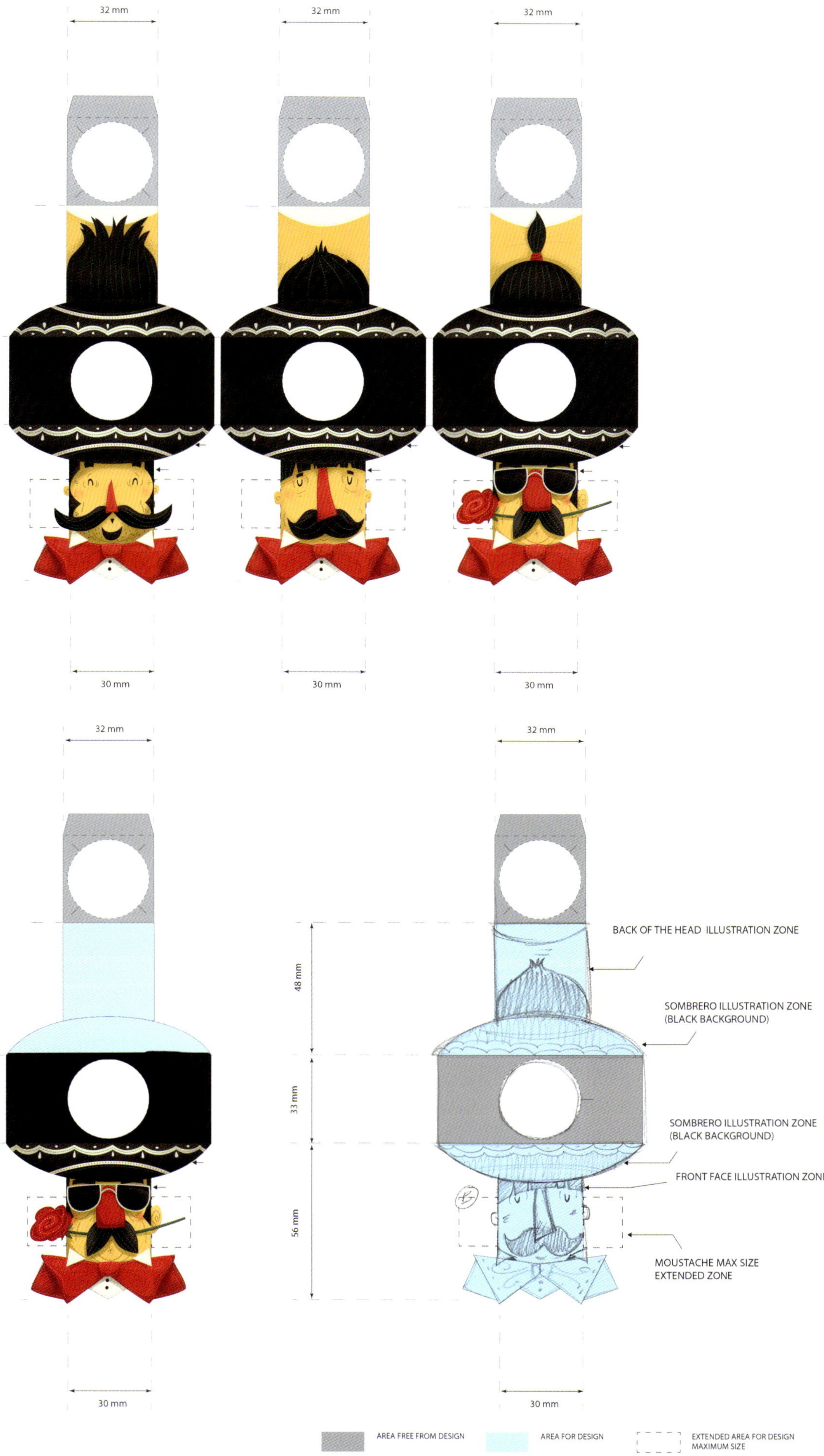
32 mm
32 mm
32 mm
30 mm
30 mm
30 mm
32 mm
32 mm
48 mm
33 mm
56 mm
BACK OF THE HEAD ILLUSTRATION ZONE
SOMBRERO ILLUSTRATION ZONE
(BLACK BACKGROUND)
SOMBRERO ILLUSTRATION ZONE
(BLACK BACKGROUND)
FRONT FACE ILLUSTRATION ZONE
MOUSTACHE MAX SIZE
EXTENDED ZONE
30 mm
30 mm
AREA FREE FROM DESIGN
AREA FOR DESIGN
EXTENDED AREA FOR DESIGN
MAXIMUM SIZE

ARDBEG ESCAPE GIFT PACK

Ardbeg has partnered with Dapy to develop a beautiful and exclusive gift box.

This elegant pack is injected in matte black ABS. It combines traditional, modern and original design. Looking like a pirate treasure chest abandoned on the sand, this coffret offers 1 bottle of Ardbeg finest scotch whisky and 2 glasses that try to escape, deforming the cover. Riveted ABS straps are screwed on the cover and base.

Opening is allowed by black PP hinges hidden inside the ABS straps. Inside, the bottle and 2 glasses are presented and held in a matte PS vac-form. Closure is done by black magnets inserted in the ABS straps.

Result is a resistant and exceptional gift.

Design Agency:
DAPY

LEFT VIEW

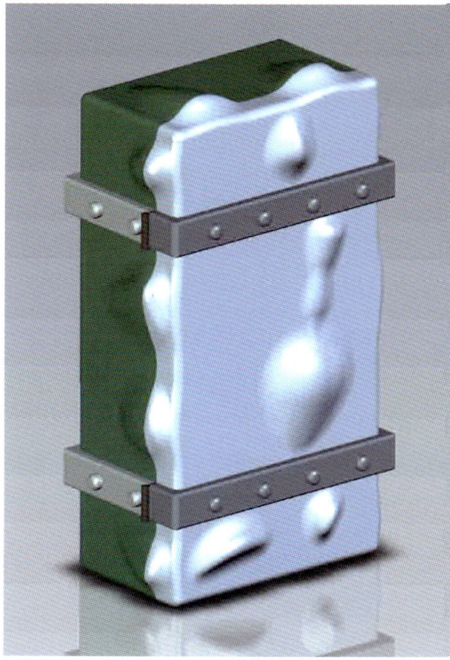

FRONT VIEW

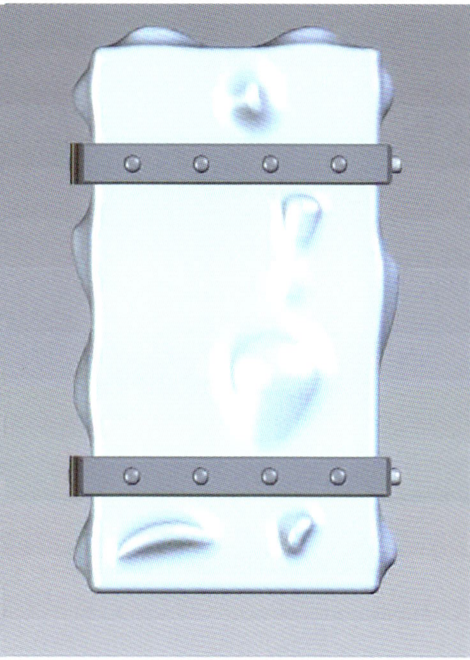

RIGHT VIEW

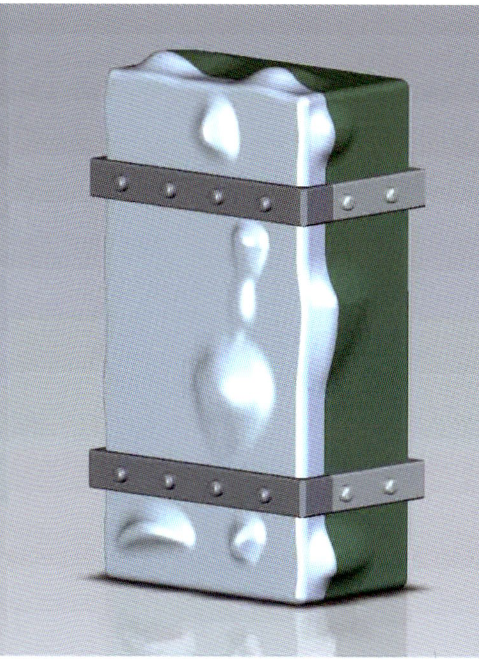

BACK VIEW

OPENED COVER

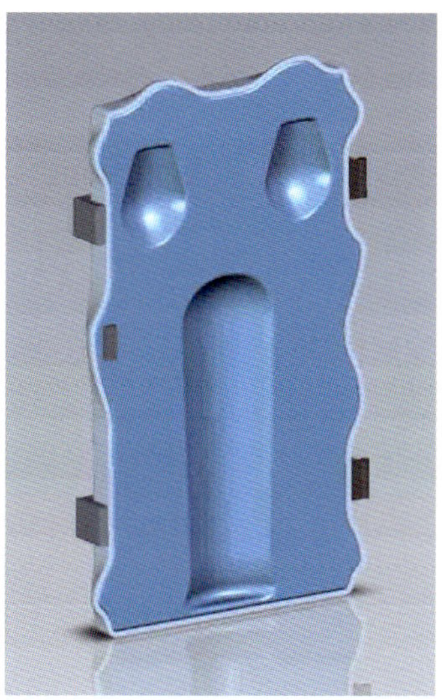

OPENED BASE

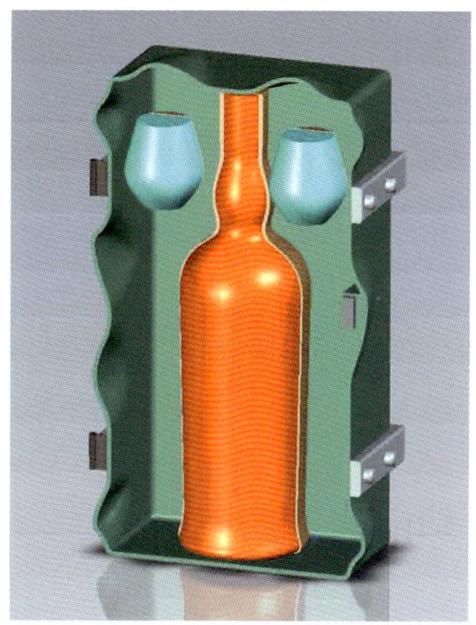

1842 RATAFIA

With a balance of styles between 19th and 21th centuries, a new identity of Ratafia 1842 is born. This is traditional liquor made from the first ratafia's recipes found in Catalonia and written in Santa Coloma de Farners town by F. Rosquellas in 1842. The client wants to update 2 labels of spirits. Commonalities were detected and we propose to make one line. It's accepted and drives him to make a 3rd spirit based on an ancient recipe from the same 1842 year found in the municipal archive. The 3 versions of Ratafia 1842 were born: Classic, Green Walnut and Eight "Porrons". Product differentiation is emphasized: ancient recipes (from 1842). Numbers "1", "8", "4" and "2" surround the bottle and this fact allows play with them rotating one or placing four side by side to see the most value of this liquor, the date of origin of the recipe: 1842. When there is only one bottle, the front ("1" number) shows the value to be the first one. A new smaller bottle with screw cap and one ink serigraphy are chosen to keep costs. Each model have different colors and typographic details. The Green Walnut and Eight "Porrons" recipes are a limited edition of 700 bottles each one for the Santa Coloma de Farners Ratafia's Festival, and were sold out with two days. That's the result of be involved in all phases of the project and to go one step further of the initial briefing.

Design Agency:
La Fonda Gràfica

Designer:
Oriol Piferrer Saló

Client:
Santa Coloma de Farners City Council & Confraria de la Ratafia

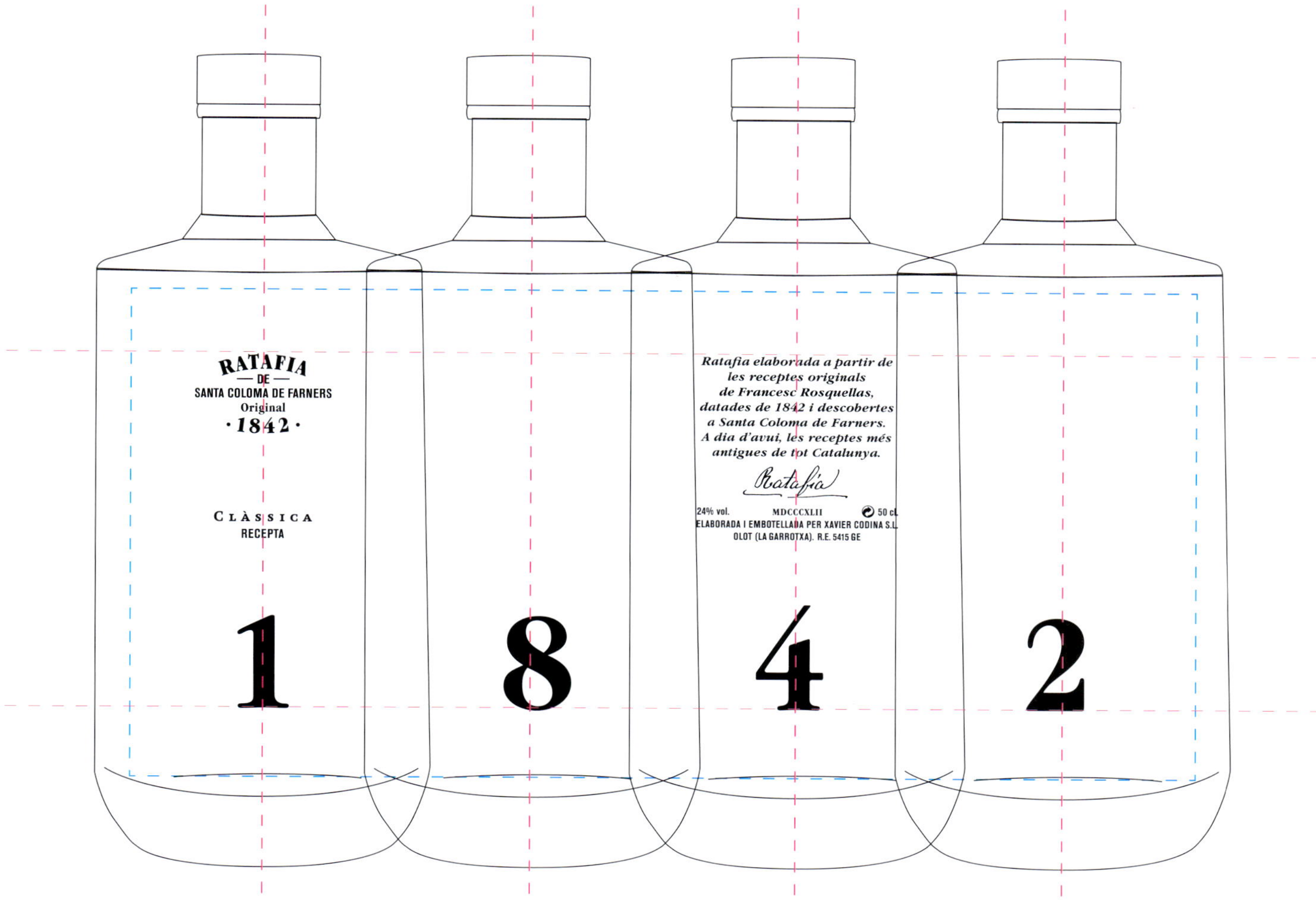
RATAFIA
DE
SANTA COLOMA DE FARNERS
Original
·1842·
CLÀSSICA
RECEPTA
1
8
Ratafia elaborada a partir de les receptes originals de Francesc Rosquellas, datades de 1842 i descobertes a Santa Coloma de Farners. A dia d'avui, les receptes més antigues de tot Catalunya.
Ratafia
24% vol.
MDCCCXLII
50 cL
ELABORADA I EMBOTELLADA PER XAVIER CODINA S.L.
OLOT (LA GARROTXA). R.E. 5415 GE
4
2

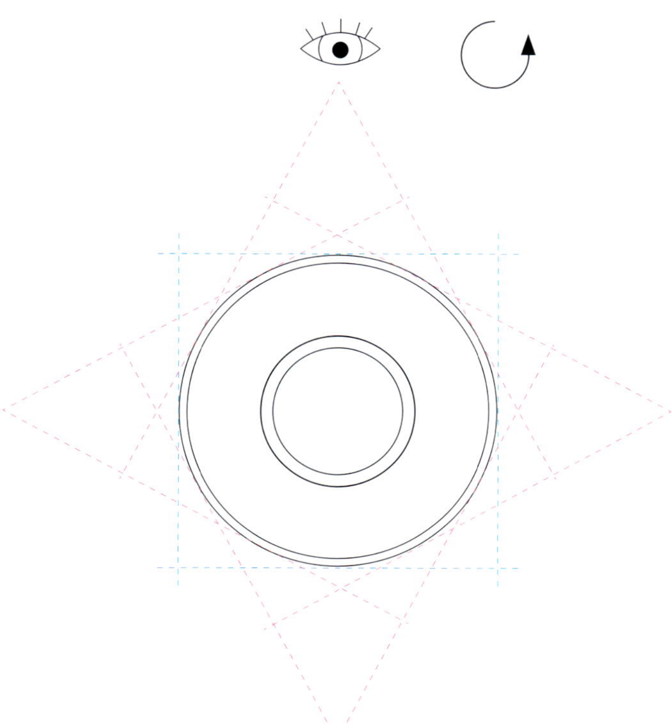

RATAFIA
DE
SANTA COLOMA DE FARNERS
Original
·1842·
CLÀSSICA
RECEPTA
1

Design Agency:
MILK Branding Professionals

Creative Director:
MANOLIS SPORIDIS

Art Director:
MANOLIS SPORIDIS

Illustrator:
MANOLIS SPORIDIS

Client:
Septem Microbrewery

SEPTEM HOBBY HOPPY, BEER SERIES

Septem Microbrewery created the Hobby Hoppy Series in order to present to beer lovers with unique, "one off" recipes which share hops as their main ingredient.

MILK Branding Professionals designed the identity and labels of the series in a way that highlights its one-off, experimental and "handmade" character. Septem ACE, the first beer of the series, won the Best New Packaging beer in the World award, in the International Beer Challenge, the most famous beer competition worldwide.

The challenge was for the overall brand identity to bring out its "handmade" and "experimental" values, avoiding the pompous style that would, in this case, reduce the sense of authenticity found in these one-off beers from a microbrewery like Septem. The series logo, each label and the overall look and feel of those, including colors, text, simplicity of typography and the quality of the paper chosen, build on the "perfectly imperfect" character of the series. Beyond winning an international prize, Septem Hobby Hoppy Series have managed to gain the respect of beer lovers in Greece, who have turned ACE and CITRA bottles into collectibles.

Septem

MONOGRAPH

Brand identity and label design for a new wine range of Gaia Winery

Monograph is a single creation crafted from a single passionate creator. The creator declares his own work to be a "Monograph" due to its unique, full and sophisticated character. This exclusive creation establishes the maker as an authority on the topic. An authority on the art of winemaking.

MILK Branding Professionals, a top-of-the range branding agency located in Athens, Greece designed from scratch a new brand for GAIA Wines. Founded in 1994, Gaia Wines is one of the most prominent Greek wine companies which owns two modern wineries located in two of the most interesting and promising P.D.O. areas of Greece, Nemea in Peloponnese and Santorini. Gaia's main aim is to present the potential of the indigenous Greek grape varieties in the wine enthusiasts worldwide. Therefore the goal for MILK Branding Professionals was to create a series of wines primarily targeted to the international markets that would convey the ideals and inimitability of the art of winery. "It was a great challenge for MILK to handle a full range project which included concept, naming and visual identity design for a distinguished winery such as GAIA."

Design Agency:
MILK Branding Professionals

Creative Director:
LEONIE YAGDJOGLOU

Art Director:
LEONIE YAGDJOGLOU

Illustrator:
LEONIE YAGDJOGLOU

Client:
GAIA WINERY

Copywriter:
ELIZABETH KATOUNTA

YÖY, ORGANIC JUICE

The main idea of this design is to transmit the high quality of a 100% organic juice directed to a target of health conscious people. Conceptually "the juice is so organic, the fruit is plucked from the tree to your mouth". The combination of naming, industrial design, and graphic design was the key to develop this packaging. The tree was the principal inspiration for the project giving birth to the name YÖY, which means tree in the indigenous language Pemon. The design of the bottle consists of small edges or vertebrae that simulate the roots of a tree and combined with the label design it evokes nature with an attractive look.

Designer:
Maria Duriana Rodriguez & Cristina Maldonado

© Master in Packaging Design ELISAVA. Barcelona, 2014.

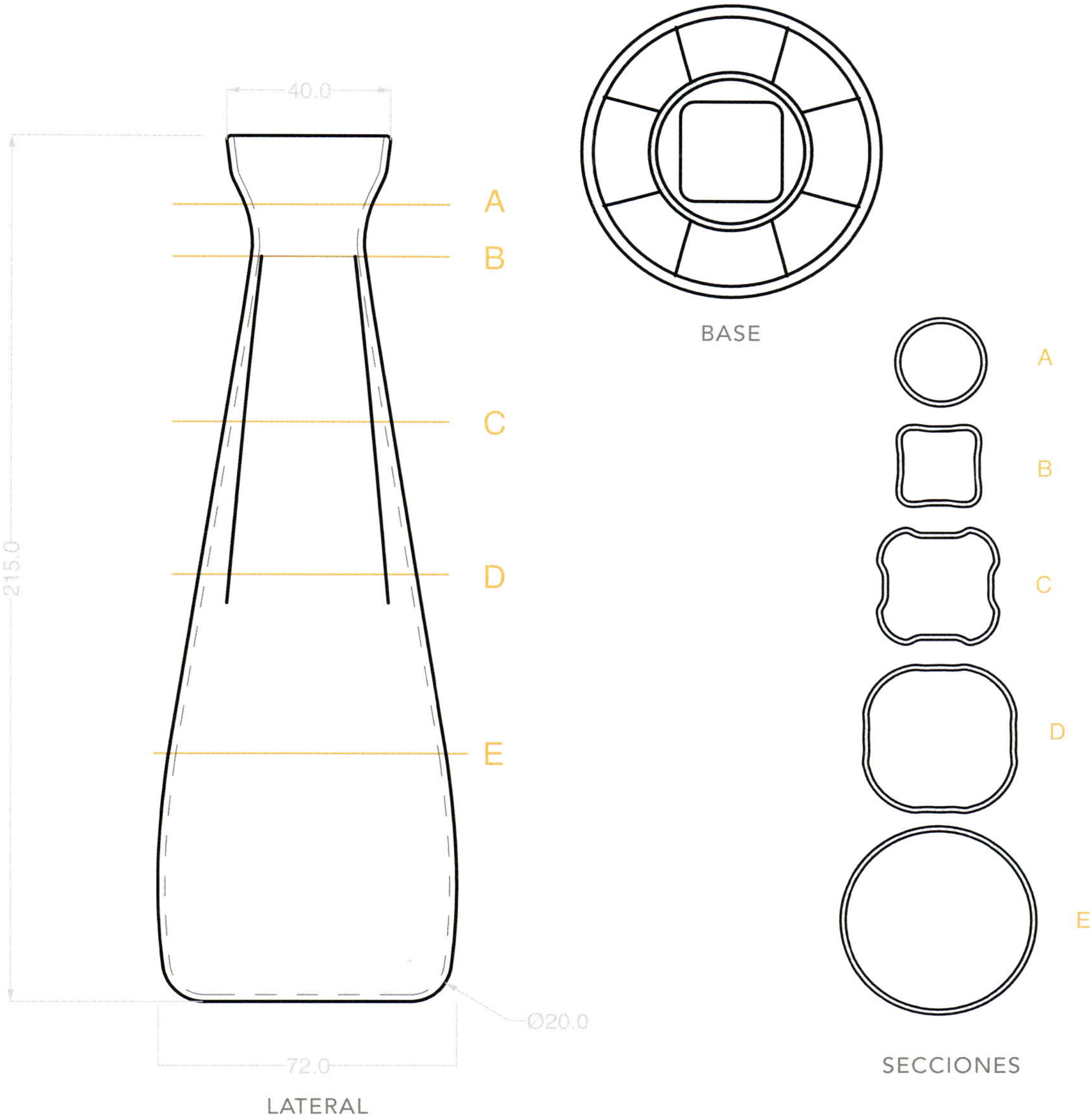
40.0
A
B
C
D
E
215.0
Ø20.0
72.0
LATERAL
BASE
A
B
C
D
E
SECCIONES

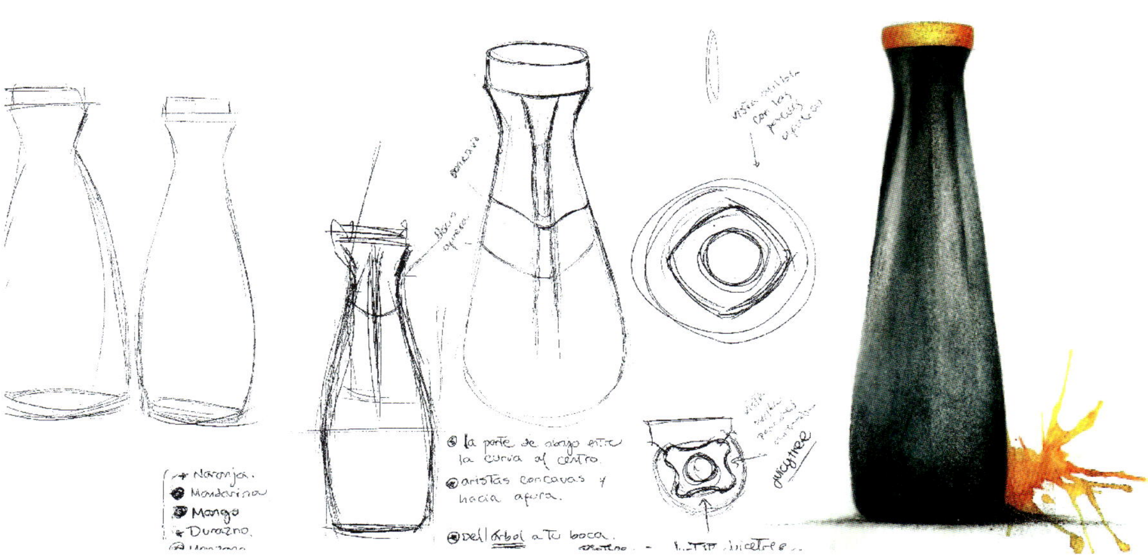
Naranja.
Mandarina
Mango
Durazno.
la parte de abajo entre la curva al centro.
aristas concavas y hacia afuera.
Del árbol a tu boca.

SA PILSEN BEER

Dressed in attractive yellow, the label is set to wow. Big and bold fonts coupled with equally awesomely big bold labels, these beer are for the strong!

Design Agency:
Lud/Co.

Designer:
Renan Vizzotto

Client:
Sa Beer

Location:
Florianópolis, Santa Catarina, Brazil

Photography:
Renan Vizzotto

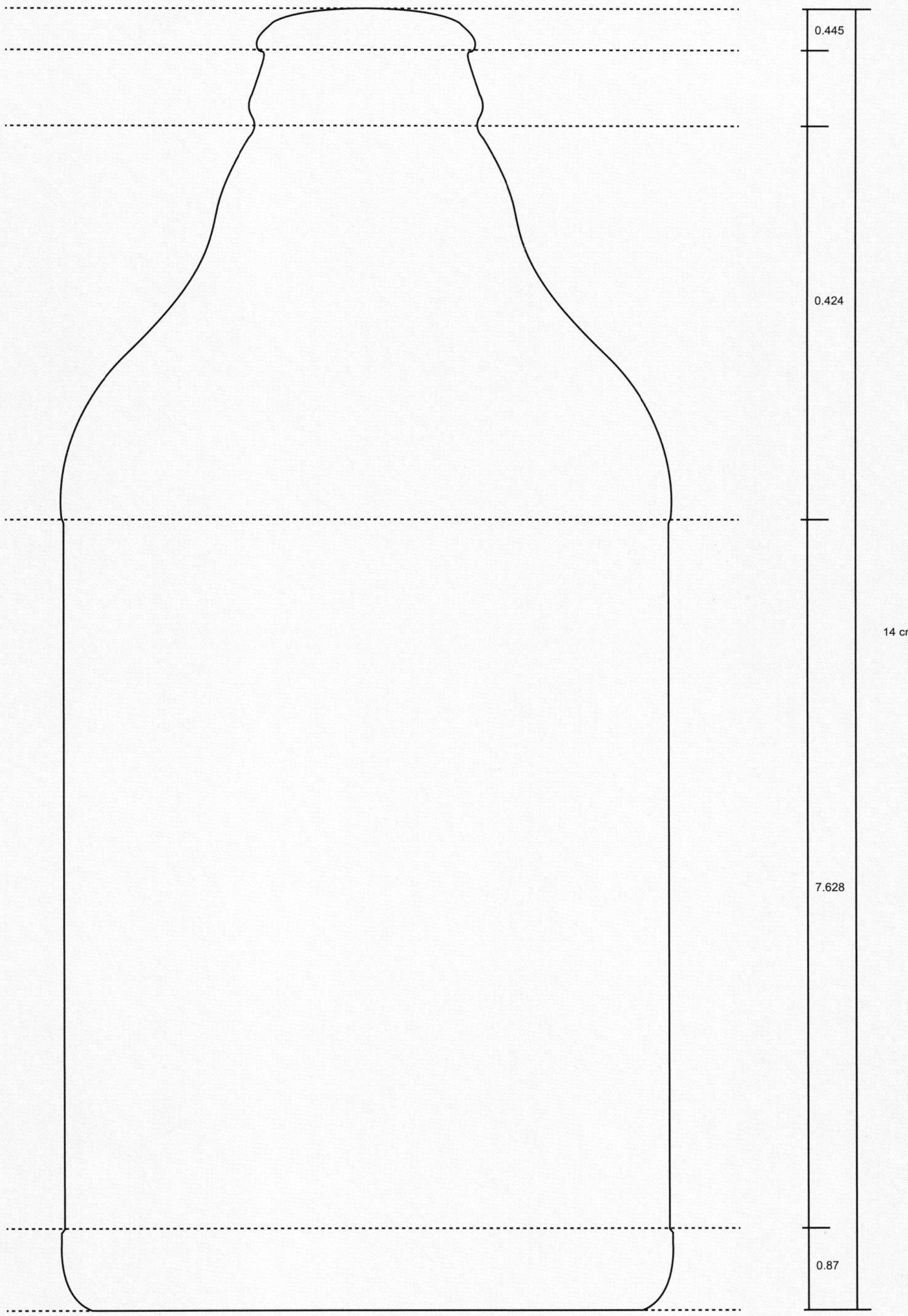
0.445
0.424
14 cm
7.628
0.87

WINE POUCH (R)EVOLUTION

Design Agency:
Reverse Innovation

Creative Director:
Mirco Onesti

Structural Designer:
Gustavo Messias

Graphic Designer:
Fiona Martin

Illustrator:
Fiona Martin

Client:
Gigante

Brand:
Rosso del Vigneto Nuovo

Photo Retouching:
Antonella Iozzolino

Photography:
Thomas Libis

How to overcome the negative image of wine sold in pouches or as a "bag in a box", which is often synonymous with poor quality? This was the question confronted by Reverse Innovation and winemakers Gigante: to develop a graphic and structural pack that appeals to a discriminating audience.

Wine bottles are intended to be touched and handled so consideration of the aesthetic, tactile and functional experience was essential. The design is a truly innovative structural pack that cleverly and elegantly reinterprets the classic Bordeaux bottle by means of an intriguing play of contrasting opaque and gloss materials. A simple black pouch is wrapped in a seductive, sensual paper sleeve that defines the elegant and intriguing shape of the product. The paper structure also ensures that the pouch is robust and self- standing at every stage of consumption.

The graphic design and fine detailing plays a crucial role in supporting the high quality positioning of the wine. The contours of the bottle shape are emphasised with gold foil while the terroir of the wine is retold through the use of blind embossing and UV varnish to reproduce the specific shape of the vine leaf which is characteristic of the area. The resulting design demonstrates how it is possible to innovate even in a segment as traditional and resistant to change as wine packaging.

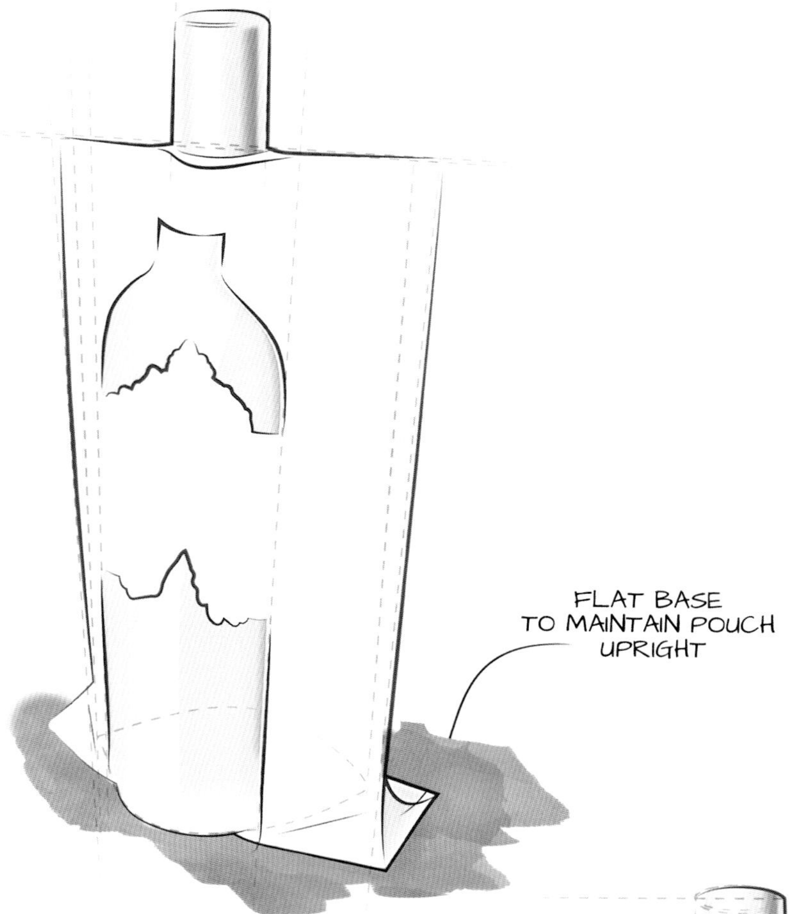
FLAT BASE
TO MAINTAIN POUCH
UPRIGHT

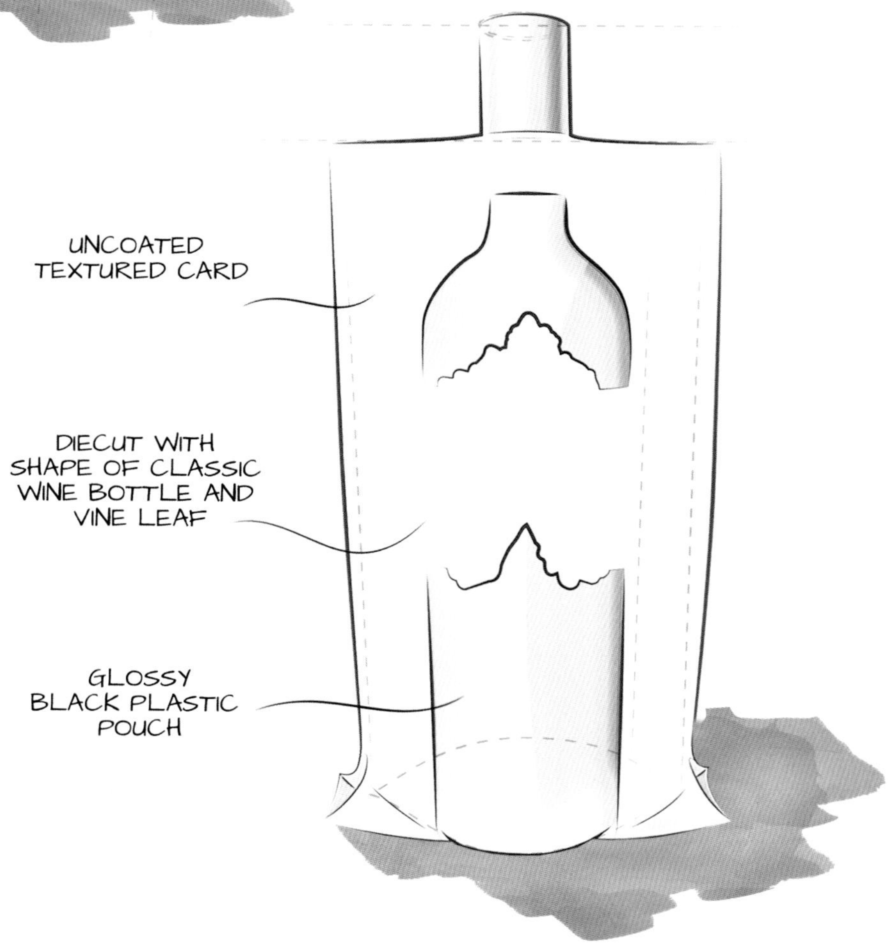
UNCOATED
TEXTURED CARD
DIECUT WITH
SHAPE OF CLASSIC
WINE BOTTLE AND
VINE LEAF
GLOSSY
BLACK PLASTIC
POUCH

SUNN VILL BRIDAL BOUQUET SERIES

For the past 40 years, we have been following the traditional brewing procedures and using modern sophisticated equipment to make high quality wine. Based on our deep care for the land and in view of the difficulty of preserving Taiwanese fruits, we now start to brew wine from fruits. In order to embody the kindness of people and fragrance of fruits for the abundant harvest from the land, we stepped out of the winery, brought the brewing techniques, and really walked into the farmers by the commitment of contract.

Borrowing the image of a Taiwanese flower, the packaging combines a flower bouquet and wine bottle together and is designed as a bottle wrapped in a one-piece flower printed film. This design not only combines beauty and utility but also solves the problem about boring display of ordinary wine bottles and turns it into a splendid flower exposition.

Design Agency:
Victor Branding Design Crop.

Designer:
Wan Ju Lee / Wei Jen Huang

Client:
SUNN VILL

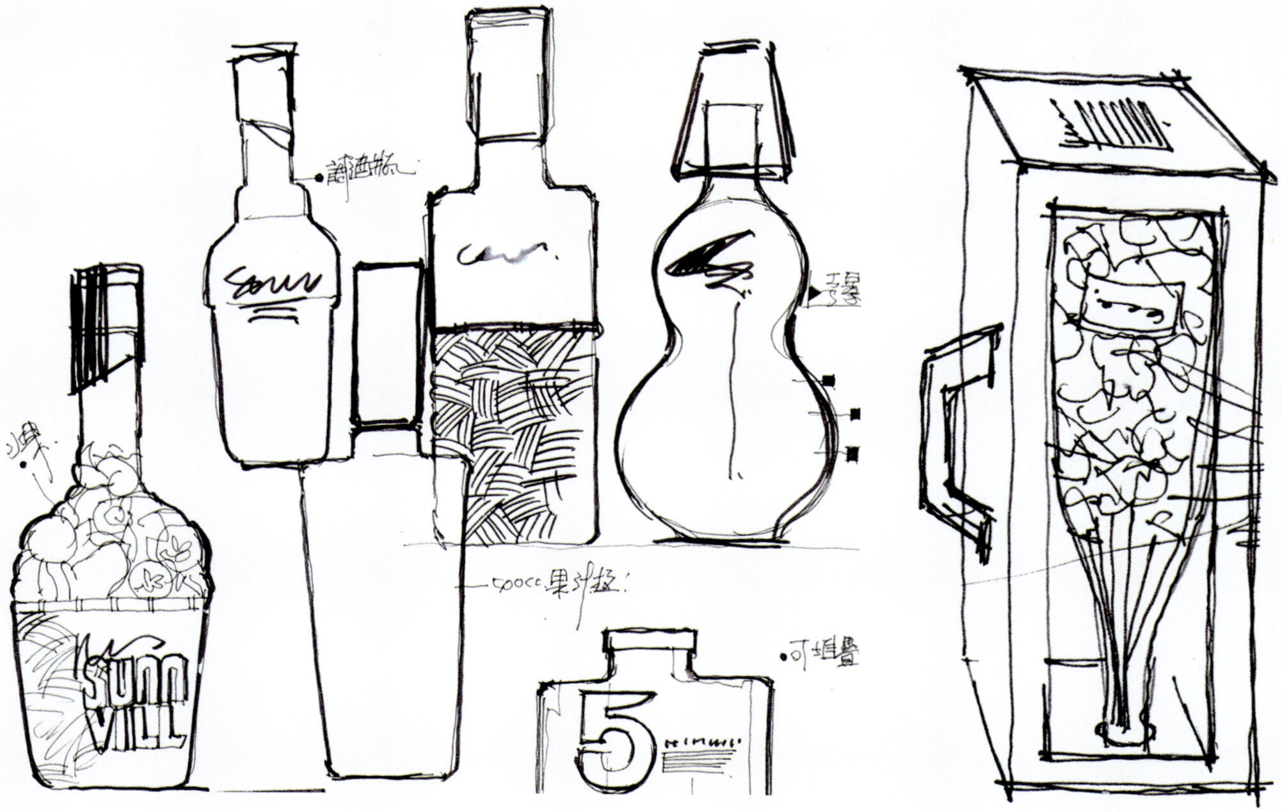
調酒瓶
可堆疊
SUNN
VILL
5

ROSE
ORCHID
CHERY

GALLO AZEITE NOVO 2013.2014

This olive oil is born from the unique and rare moment of the first olives harvested.

The design should reflect the taste of a new experience and create differentiation in the market shelf. The dynamic graphics and vibrant colours appeal to younger consumers.

Painted in oil canvas, the illustration was then photographed and digitally composed.

The packaging was designed to the maximum expose of the bottle and to support a small plate.

Every year, Gallo Azeite Novo introduces a new packaging with the purpose of being collectible.

Design Agency:
RMAC BBDO Portugal

Designer:
Alexandre Mendes

Client:
Gallo Worldwide

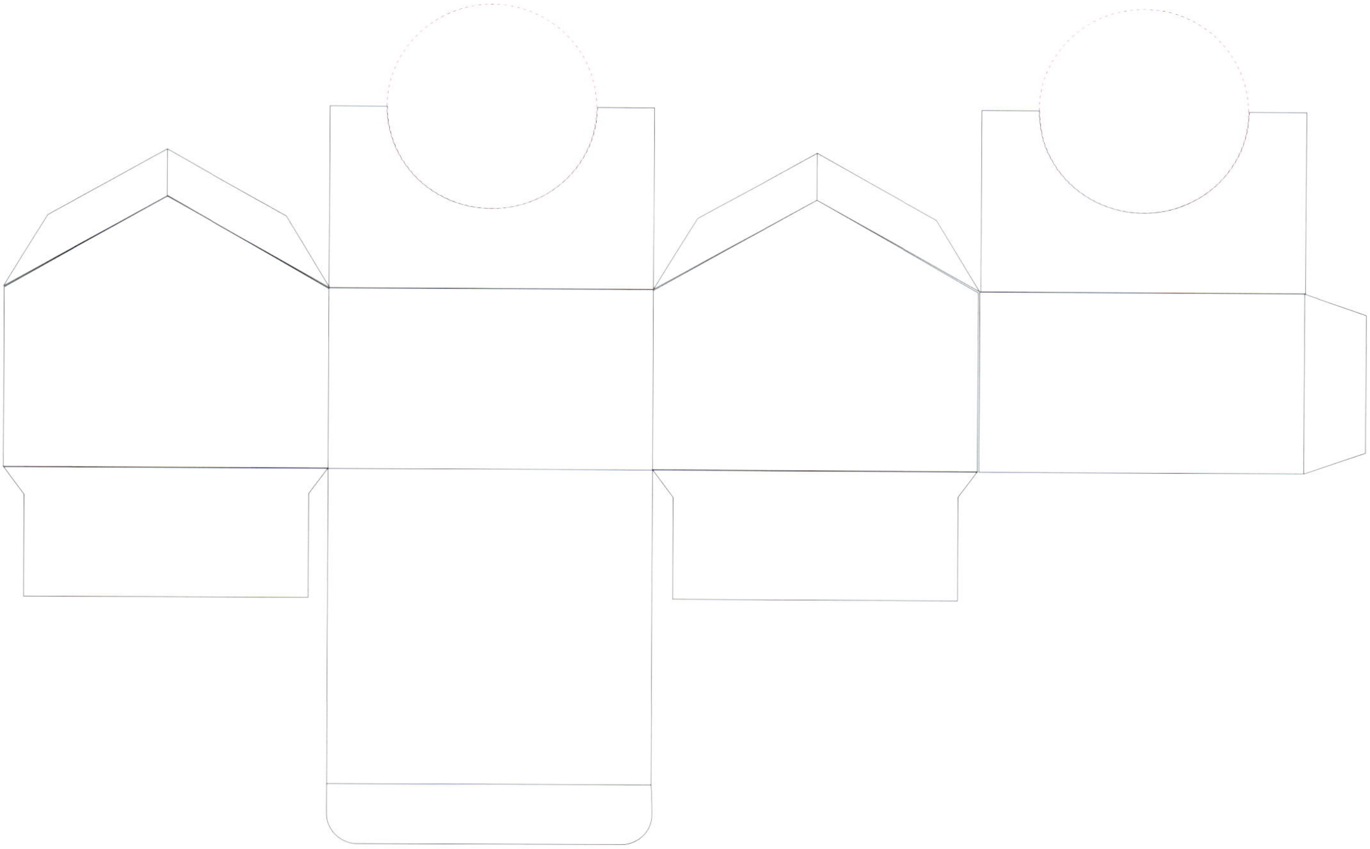

GALLO
500ml ℮
GALLO

EDIÇÃO LIMITADA

ICE CRUSH

The Ice Crush is a packaging for the Belvedere premium Vodka. The design is inspired by natural ice blocks and ice fields. At home, the packaging has a second life and it can be used as an ice mould to make ice blocks and recreate the ritual of Belvedere Martini.

Design Agency:
5.5 designstudio

Client:
Belvedere vodka

Photography:
5.5 designstudio

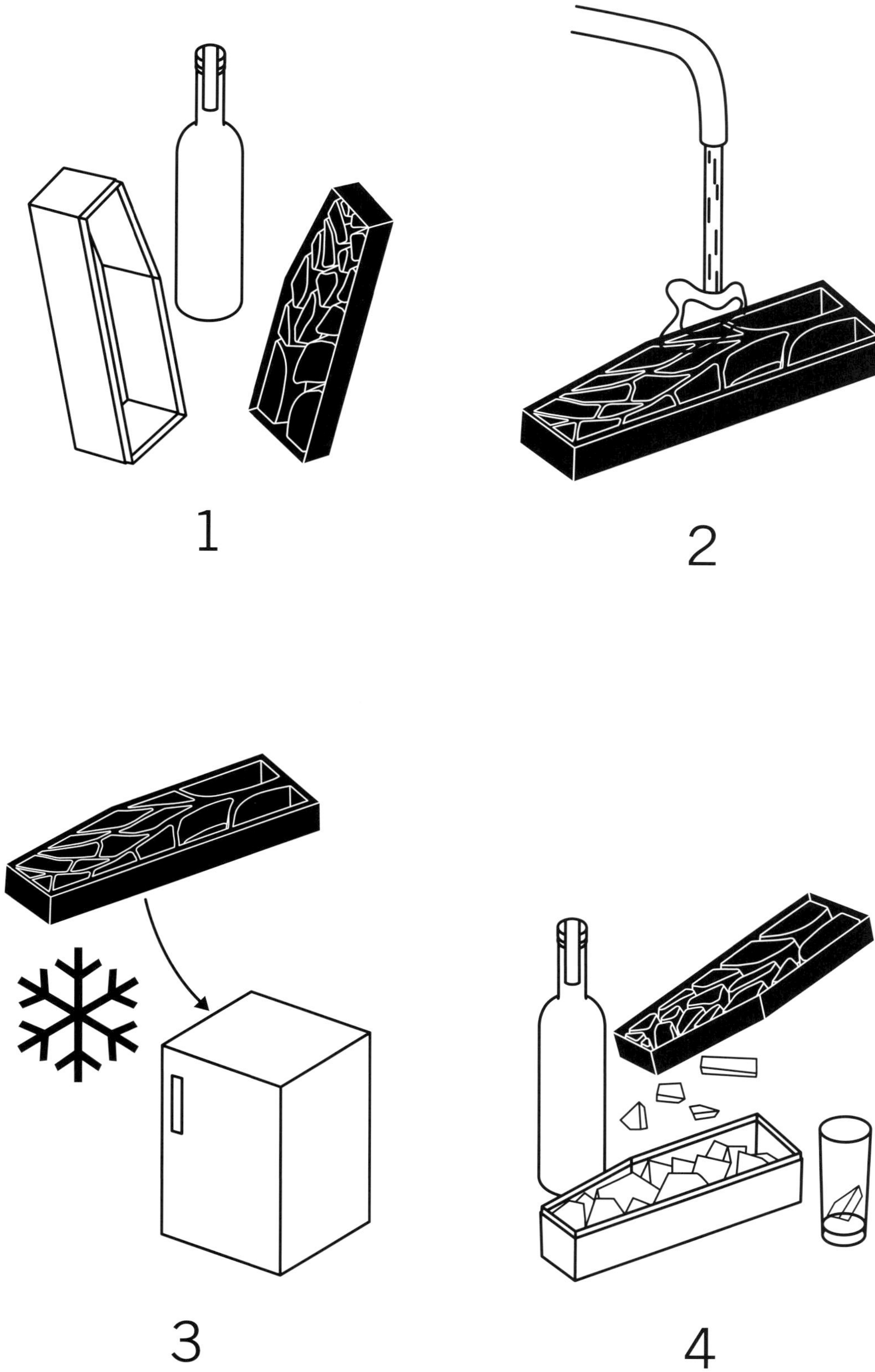
1
2
3
4

GOLD MINE BEER

The first offbeat work on Russian beer market. The PET packaging imitates the original beer glass and the lid approaches beer foam. This work was marked on many design competitions and won Pentawards Gold 2014.

Design Agency:
Svoe Mnenie

Creative Director:
Andrey Kugaevskikh

Client:
Efes Rus

Photography:
Mikhail Samoshkin

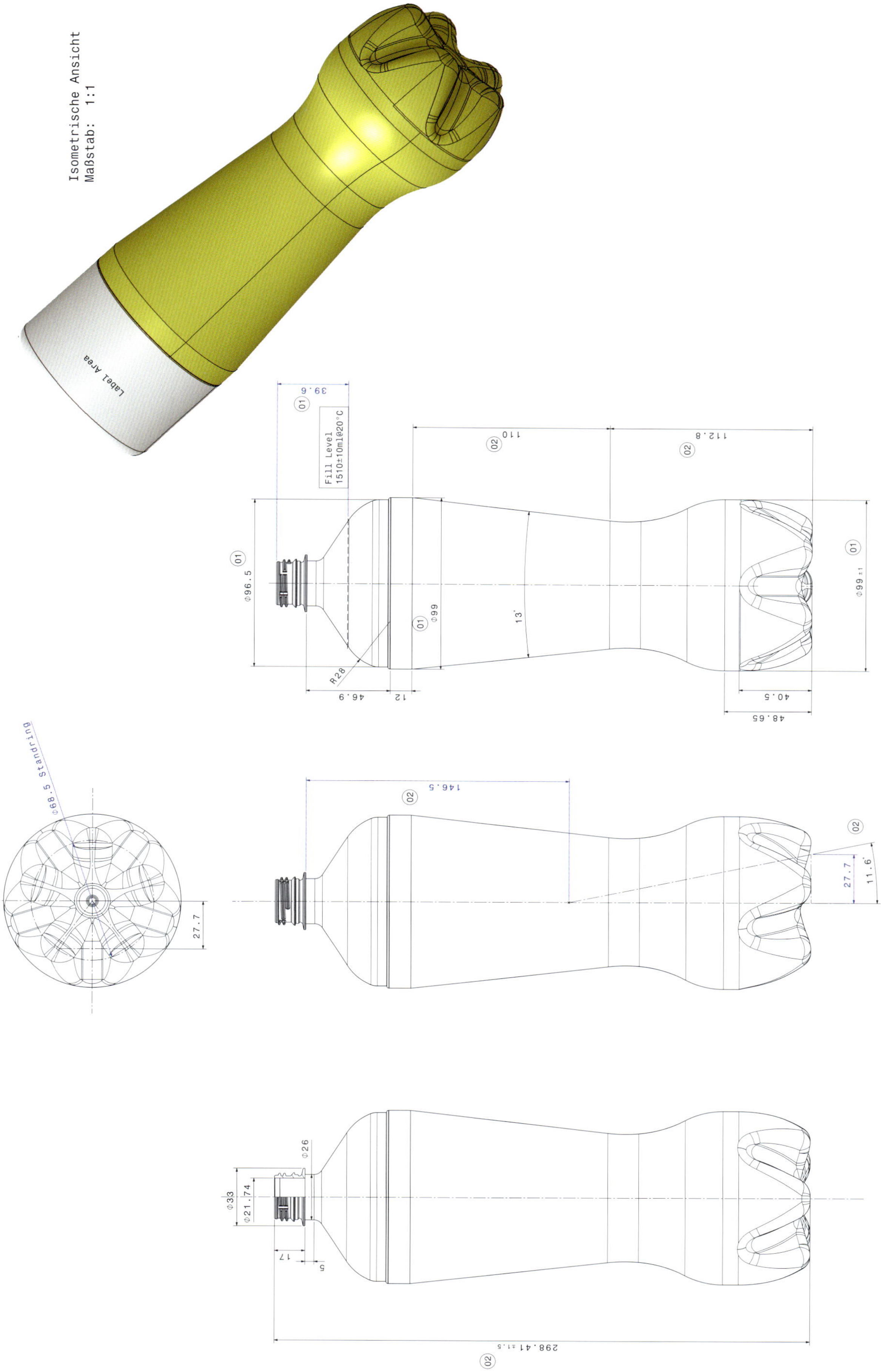
Isometrische Ansicht
Maßstab: 1:1
Label Area
39.6
Fill Level
1510±10ml@20°C
110
112.8
Φ96.5
Φ99
13°
R28
46.9
12
Φ99±1
40.5
48.65
Φ68.5 Standring
146.5
27.7
11.6°
Φ26
Φ33
Φ21.74
17
5
298.41±1.5

MISCELLANEOUS

Designer:
Maria Ponomareva

Client:
BHSAD, Moscow

ARTEA

Tea set, in hand-made style designed for creative persons.

The consumer has the choice to make the tea they want which coincides with their mood and situation.

They can experiment, control and change the tea to their taste.

This project was carried out while studying at British Higher School of Arts and Design in Moscow on the course "visual communication". The project had been developed: a brand platform, brand-name, packaging and advertising in the press.

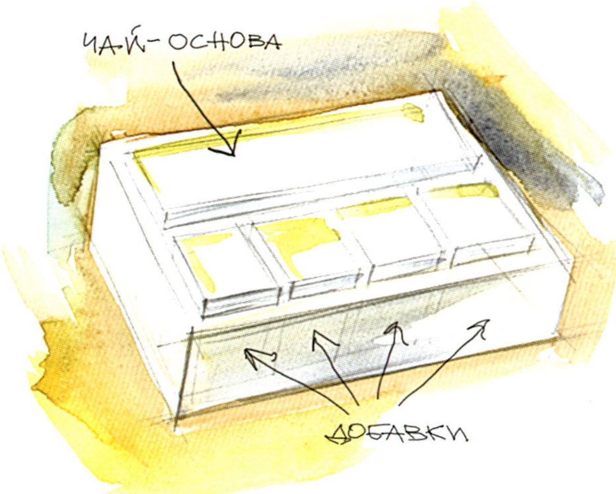
ЧАЙ-ОСНОВА
ДОБАВКИ

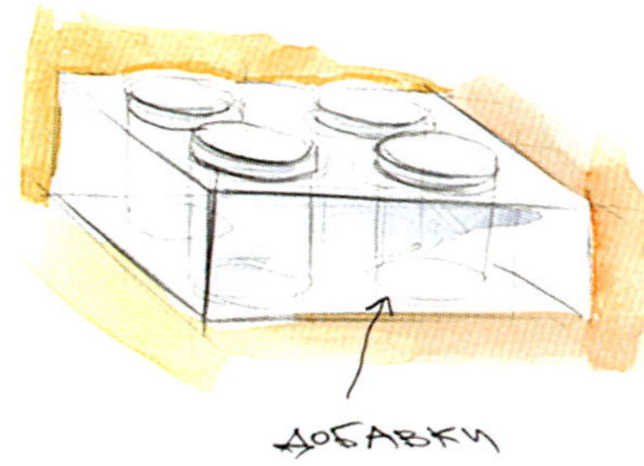
ДОБАВКИ

ЧАЙ-ОСНОВА

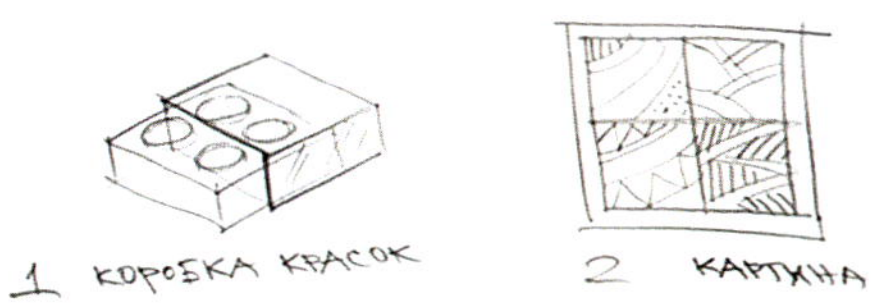
1 КОРОБКА КРАСОК
2 КАРТИНА

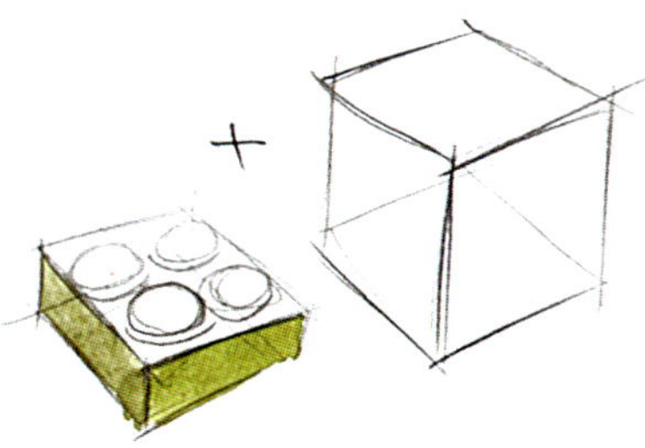

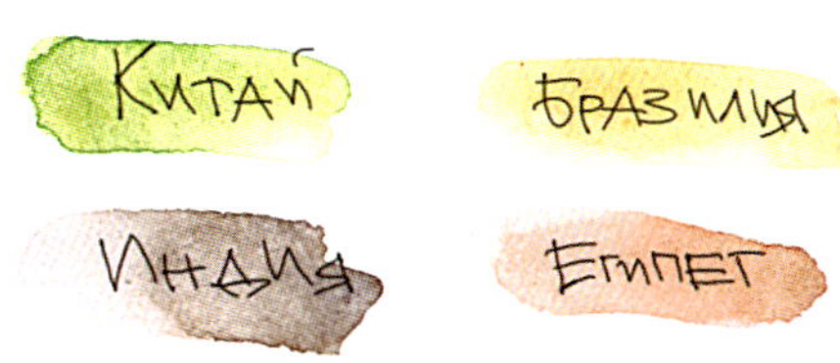
Китай
Бразилия
Индия
Египет

artea
зеленый чай
календула
тмин
хризантема
гвоздика
artea
календула
для чая
китайская
палитра
тмин
для чая
китайская
палитра
хризантема
для чая
китайская
палитра
гвоздика
для чая
китайская
палитра

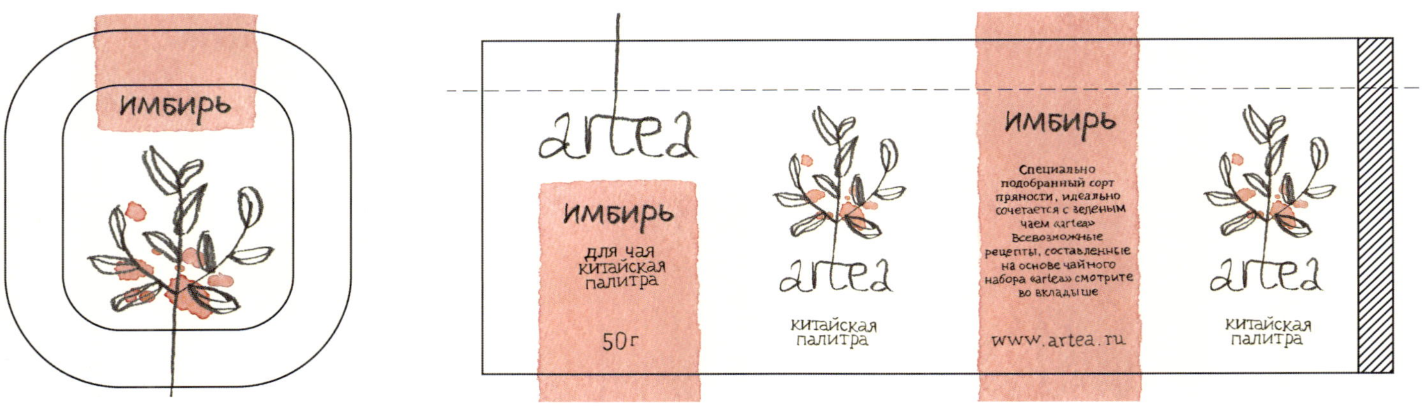
имбирь
artea
имбирь
для чая
китайская
палитра
50г
artea
китайская
палитра
имбирь
Специально подобранный сорт пряности, идеально сочетается с зеленым чаем «artea»
Всевозможные рецепты, составленные на основе чайного набора «artea» смотрите во вкладыше
www.artea.ru
artea
китайская
палитра

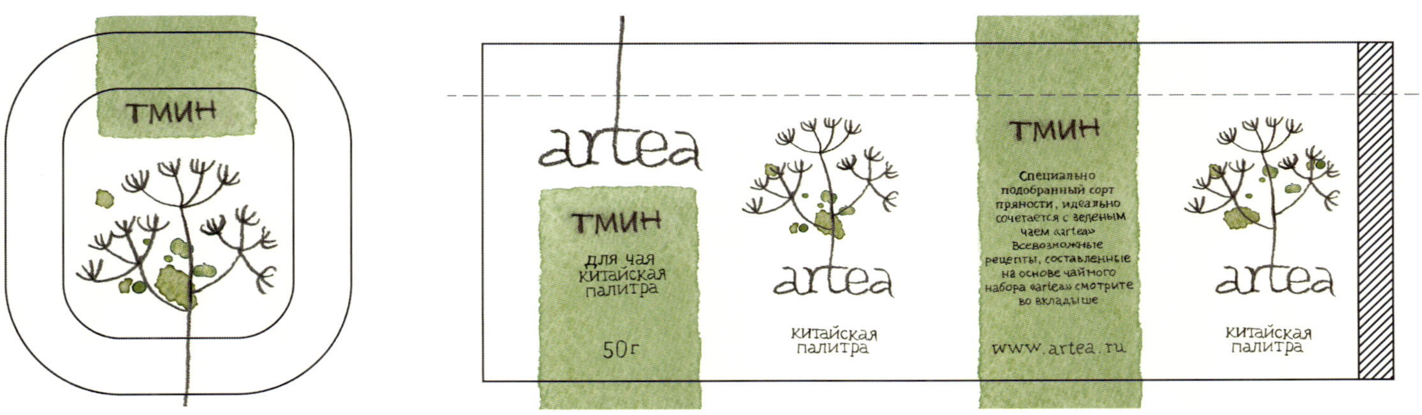
тмин
artea
тмин
для чая
китайская
палитра
50г
artea
китайская
палитра
тмин
Специально подобранный сорт пряности, идеально сочетается с зеленым чаем «artea»
Всевозможные рецепты, составленные на основе чайного набора «artea» смотрите во вкладыше
www.artea.ru
artea
китайская
палитра

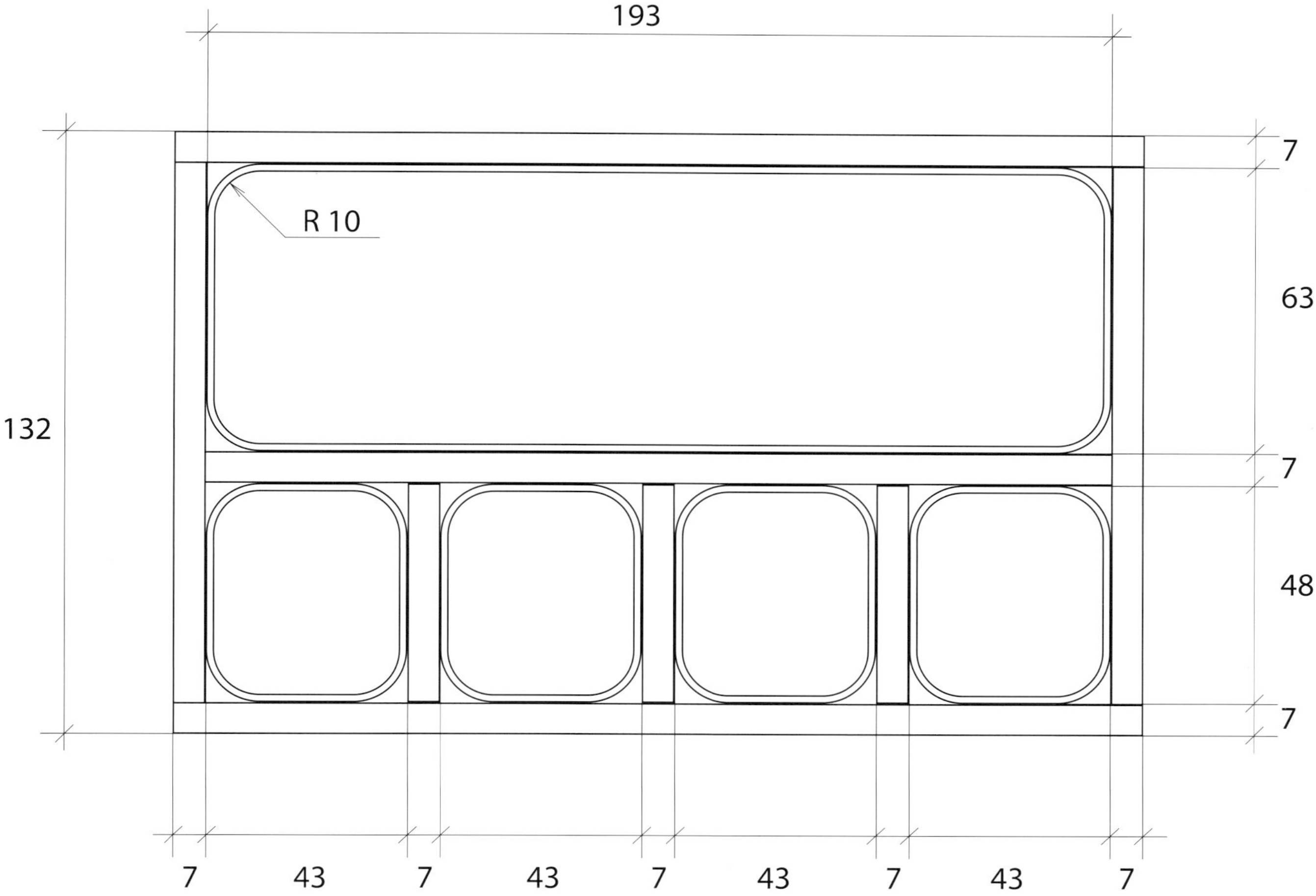
193
R 10
132
7
63
7
48
7
7 43 7 43 7 43 7 43 7

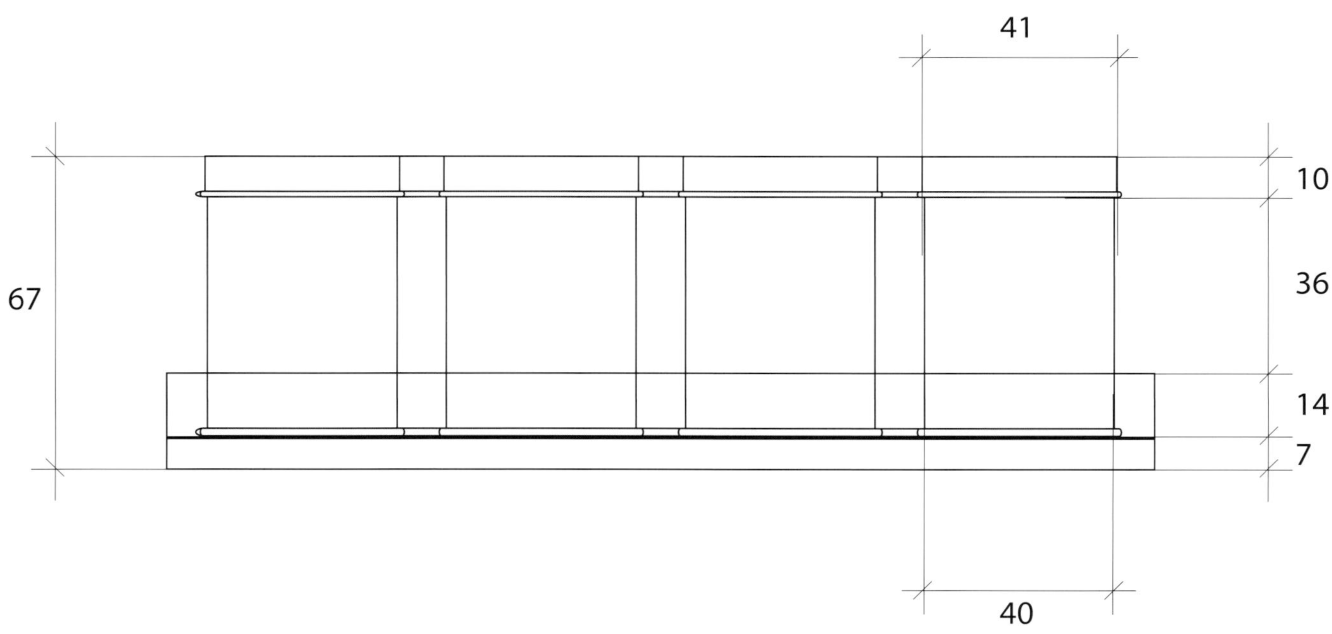
41
67
10
36
14
7
40

DOG'S DINNER

Design Agency:
Afterhours

Designer:
Moyra Casey, Kelly Bennett and Chris McDonald

Conceptual product idea, identity and packaging for a range of chilled ready meals and ambient snacks for man's best friend.

Conceived to demonstrate challenger thinking and sector innovation.

Paralleling their owners' favourite recipes but with the right nutritional balance for dogs, these mixed up dishes would look like a bit of a dog's dinner to us but have all the ingredients of a delicious meal to them.

To capture this theme of a dog's perspective on taste, the identity and structural packaging would centre around the slipper for dinner and the newspaper for breakfast-two staples of family life that hold an unaccountable attraction to dogs.

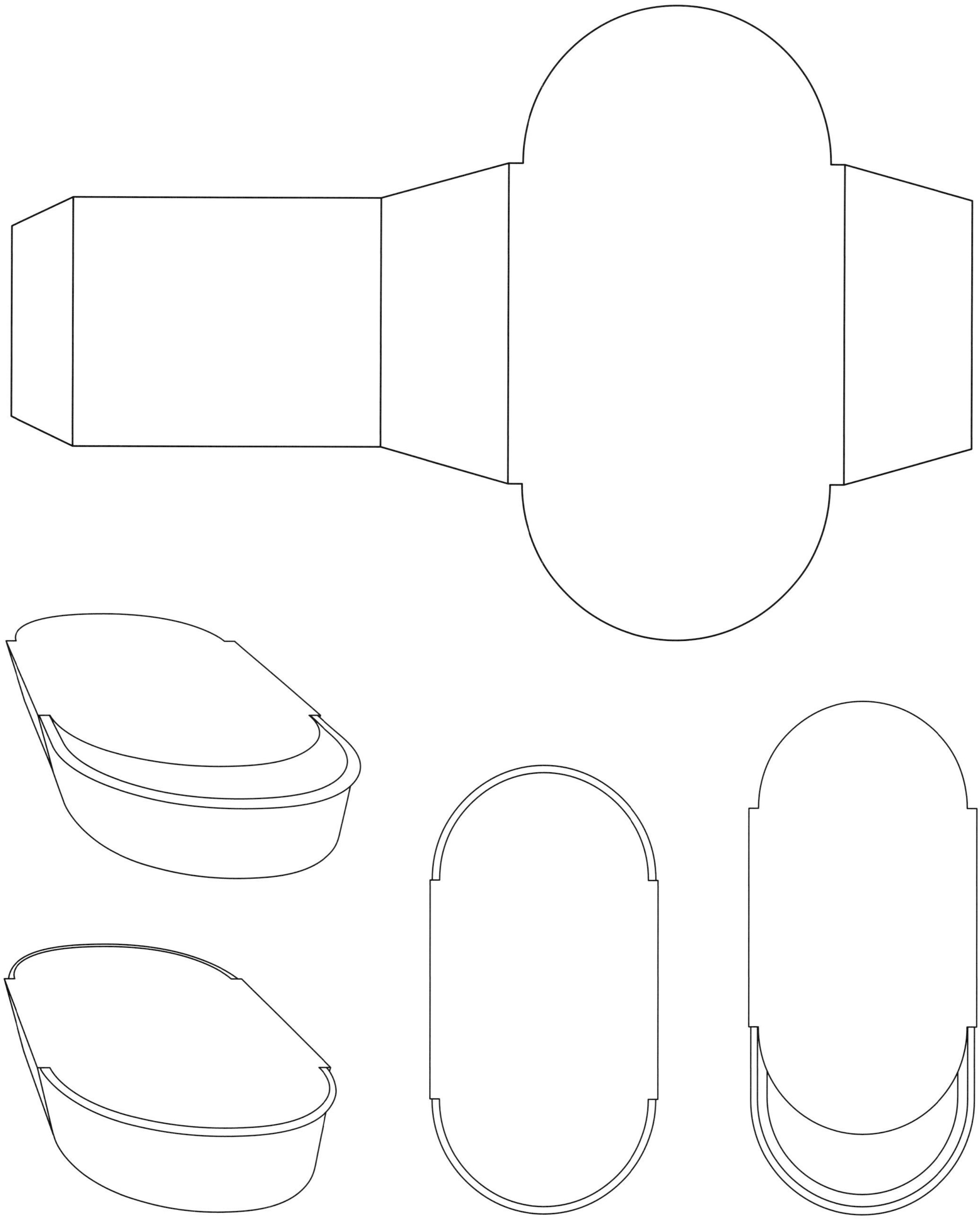

DOG's BREAKFAST
Best Before 24th July 2014 dogs breakfast.co.uk issue no 20198
Only £2.45
PAPERBOY QUITS!
"That Yorkie is lethal"claims terrified boy
OATY SENSATIONS!
Crunchy oaty bites packed with vitamins and minerals to get your day off to a flying start
DOG's BREAKFAST
Best Before 24th July 2014 dogs breakfast.co.uk issue no 20137
Only £2.45
Laundry basket raided
Angry calls for a more secure lid as thief makes off with three socks and a pair of lacy M&S knickers.
CHEESY BITES!
Crunchy, cheesy biscuits packed with vitamins and minerals to get your day off to a flying start

DOG's
KFAST
Best Before 24th July 2014 dogs breakfast.co.uk issue no 2013
Only £2.45
Laundry basket raided
Angry calls for a more secure lid as thief makes off with three socks and a pair of lacy M&S knickers.
CHEESY BITES!
Crunchy, cheesy biscuits packed with vitamins and minerals to get your day off to a flying start

Design Agency:
Geometry Global Frankfurt

Designer:
Maik Hofmann, Felix Dürichen, Stefan Hecht

Client:
Fedrigoni Germany

THE PAPER SKIN

A unique design idea telling a unique story.

From packaging to Leica's Iconic banderole: The Limited Edition Fedrigoni turned Second Skin into first skin.

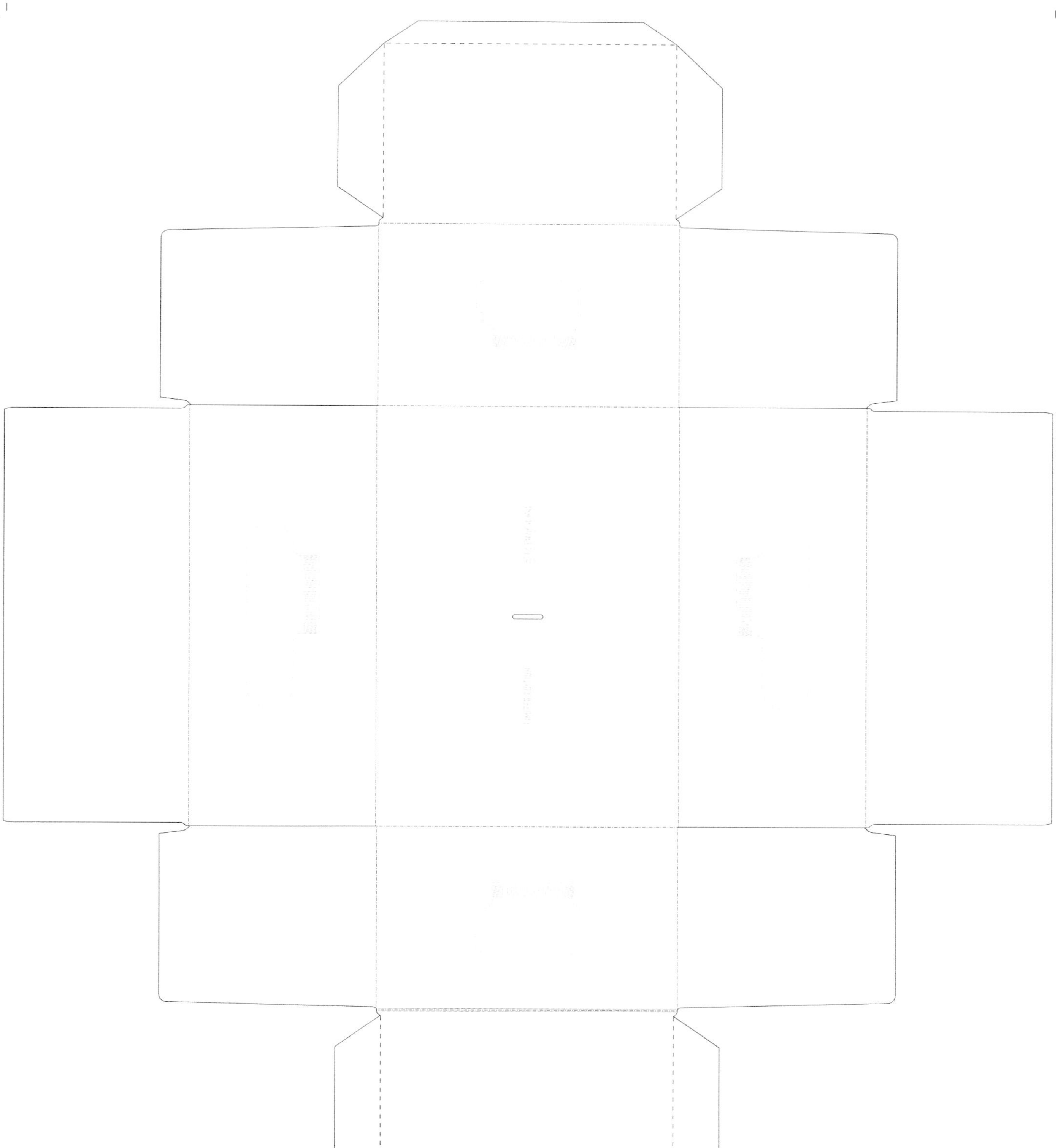

Fedrigoni's "Constellation Jade" is a packaging paper with extraordinary material properties – it gives top, premium products a unique second skin. Geometry Global worked with Fedrigoni to create excitement around a sample book for packaging paper.

To tell a craft story like no other, it required a brand like no other. So why not have it told by a brand like no other? Geometry Global and Fedrigoni partnered with Leica X2.

To address Constellation Jade's assets a sample book was designed in the guise of a posh Leica X2 packaging. But not just any Leica X2. A limited edition was created by committing a design sacrilege: Leica's iconic leather banderole was officially replaced with Constellation Jade.

A subtle design idea had turned a luxury brand into an ambassador for a low interest product, created enormous buzz amongst a design minded target, and, finally, sold paper sample books for 3,900 EUR each.

16.12.13 13:33

131216_fed_buch_2ndskinbecomes1stskin_pfade_rz.indd 1

WELKOOP BIRD BOXES

Until recently these animal shelters were "tucked-away" in cardboard boxes. Because the bird-boxes are looking pretty themselves, we have chosen for a different way of packaging. The products are completely visible and each one has its own label. At a glance you will know which accommodation suits which animal, by depicting the animal and let it have interaction with the box. The colourful assortment invites to give such a residence as a gift to yourself or someone else.

Design Agency:
Millford

Designer:
Jobert van de Bovenkamp

Client:
Agri Retail

Brand:
Welkoop

BoerenBond
welKOOP
BIJENHUIS
Bijen helpen de bloemen en planten in uw tuin te bevruchten.
• NESTGELEGENHEID VOOR VERSCHILLENDE SOORTEN BIJEN
• GEMAAKT VAN HOLLE BAMBOESTOKKEN
Made in China
BoerenBond/Welkoop, Ede
www.boerenbond-welkoop.nl
BIJENHUIS
Europa telt naast de honingbij vele honderden bijensoorten. Deze zelden stekende wilde bijen helpen de bloemen en planten in uw tuin te bevruchten. Door de achteruitgang in flora en nestgelegenheden worden vele soorten bedreigd.
INFORMATIE
Dit bijenhuis biedt een uitstekende nestgelegenheid voor verschillende soorten bijen. Een vrouwtjesbij maakt in een holle bamboestok verschillende cellen met daarin een eitje en wat pollen en nektar voor de larf.
OPHANGEN
Hang dit bijenhuis op een zonnige, op het zuiden gerichte plek.

BoerenBond
welKOOP
EEKHOORN VOEDERKAST
Met deze eekhoornvoederkast voorkomt u dat eekhoorns de vogelvoederhuizen plunderen en kunt u genieten van hoe de eekhoorns zichzelf bedienen met voedsel.
• TE VULLEN MET NOTEN
• SCHARNIERBAAR DAKJE DOOR EEKHOORN ZELF TE OPENEN
• VOORKOMT HET PLUNDEREN VAN VOGELVOEDERHUIZEN
Made in China
BoerenBond/Welkoop, Ede
www.boerenbond-welkoop.nl
EEKHOORNVOEDERKAST
Eekhoorns zijn echte acrobaten in bomen en struiken en gaan slim te werk bij het bemachtigen van voedsel. Op het menu van de eekhoorn vind je bijvoorbeeld pinda's, walnoten, hazelnoten, zonnebloempitten.
INFORMATIE
Tot spijt van vele vogelliefhebbers beschadigen of plunderen eekhoorns vaak de vogelvoederhuizen. Met deze eekhoorn-voederkast voorkomt u dit probleem en kunt u tevens met veel plezier genieten van hoe de eekhoorns zichzelf bedienen met het rijke aanbod van noten. In korte tijd hebben ze ontdekt hoe ze het dakje van de voederkast kunnen openen om bij het voer te komen. Ze zullen het ter plekke opeten of verstoppen het in de tuin om een voorraad op te bouwen voor de winter. Bovendien laten deze vlugge dieren zich eenvoudiger observeren wanneer ze aan het eten zijn.

BoerenBond
welKOOP
VOGELHUIS PIMPELMEES
VOGELHUIS PIMPELMEES
OPHANGEN
ONDERHOUD

BoerenBond
welKOOP
VOGELHUIS ROODBORST
VOGELHUIS ROODBORST
OPHANGEN
ONDERHOUD

BoerenBond
welKOOP
EEKHOORNNESTKAST
EEKHOORNNESTKAST
TIP
OPHANGEN

DOLCIARIA SERIO – SPECIAL EDITION FOR ANGELINI DESIGN

Angelini Design celebrates Christmas and the New Year with its clients and friends with the gift of a unique assortment of the best torroncini nougats from Dolciaria Serio, since 1926 a renown artisanal patisserie in San Marco dei Cavoti, in Benevento, an inland province of the Campania region.

The nougat of Benevento, whose tradition Serio continues, represents and updates, is a symbol of a land and its people, an expression of their craft, genuineness and recipes that date back to the ancient Romans. The choice of ingredients, the preparation and the cooking are in fact done with the same care and passion of those times.

Design Agency:
Angelini Design

For this special occasion the product is packaged in a black box, closed with metallic staples and string of the same color, underlining its elegance and refinement. The delicacies typical of San Marco dei Cavoti are here reinterpreted by the image and text "Torroncini di Angelini Design", printed in shiny silk-screen ink. A style of Futurist rigor, a nervous and energetic triumph of ruling pen and angles.

TORRONCINI
ANGELINI
design
TORRONCINI
TORRONCINI

BIGELOW TEA REBRAND

The work is for the 68th anniversary for the Connecticut based specialty tea company Bigelow Tea Co. To commemorate this landmark, I reinvent the Bigelow brand through an innovative product delivery system concept, introduced in a high-end gift set package and identity.

Due to Bigelow's wide target market of 18-60 year olds, I had to create a packaging concept that is fresh and modern, yet classic. The innovative tea concept I developed was Bigelow's signature teas in the form of a dissolvable powder, compacted into a sphere shape. These easy to use 'tea pods' are individually wrapped in colored cubes to represent the three flavors included in the gift set: Constant Comment, Darjeeling and Lemon Lift. I expanded on these diamond-rotated cubes in the gift box construction, resulting in a pyramid inspired structure that is modern and unique compared to the competition.

Designer:
Brielle Wilson

Client:
Concept for Bigelow Tea Co.

Photography:
Brielle Wilson

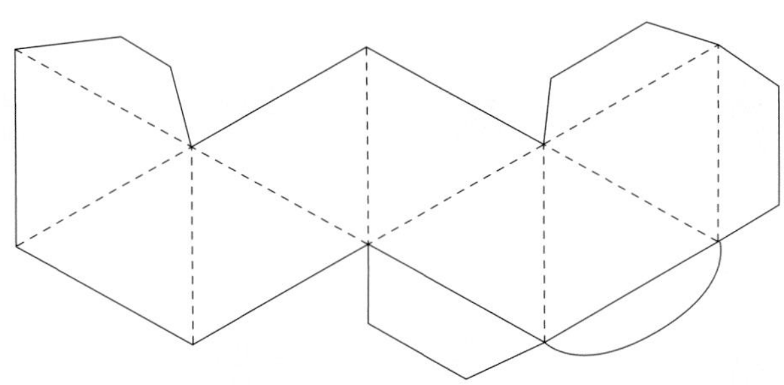

CHA TZU TANG

Seeking balance between nature and human, we long for compliance with the Mother Earth. Cha Tzu Tang is a pioneering green brand from Taiwan, generating harmony between nature and life and hoping our aromatic tea seeds can slow down our daily paste.

From interior to exterior, we carefully take care of the valuable tea seed which only produce once a year; Completely, we bring this mellowness into daily life of human being. We also insist on the idea of taking from the nature and return it to nature. And make all things involved in tea seed continuously recycle and enhance nature beauty. Inherit the artisan spirit of old masters in Cha

Design Agency:
Victor Branding Design Crop.

Illustrator:
SHIU, RUEI JR

Client:
Cha Tzu Tang

800ml 500ml 300ml

Tzu Tang, the growth of tea seed and production process are reviewed and innovated, the limit of old technology is improved. The tea seed oil produced by repression and refining are mixed with tea glycosides according to different proportion, human-oriented four series related to oil, hair, body and family are developed, integrated with the plain and sincere emotion of this land, enable us to feel the nature and purity of tea seed generated in Taiwan mountain and forests from inside to outside when we enjoy the breath daily.

cha
tzu
tang

MEDICINE PACKAGE

This was the school project, when the task was to design packing for 3 different medicines. My goal was to create a clean and strong identity that reflects those diseases. My aim was to make something different. I would like to apply special methods, such as hologram sticker, embossing and use different kind of papers. Beginning I used the microscopic pictures of the diseases, as a sample. Hereinafter I created a frame, which forms are specific for the diseases come to being by combination of different tones, colors and lines. It was an important point of view, that customer recognise easily the medicines, however do not has a bad feeling to use it. Just on the contrary. I would like to reach that customers do not afraid to use it any times if it is necessary. These medicines are for different illnesses: Curiosin -for treating scars and wounds, Cariprazine - for treating neurogical disorders, Oculo plus - for improving the eyesight.

Designer:
Attila Ács

Creative Director
Attila Ács

Client
School project (Hungarian University of Fine Arts)

MONOCROME STRUCTURE_

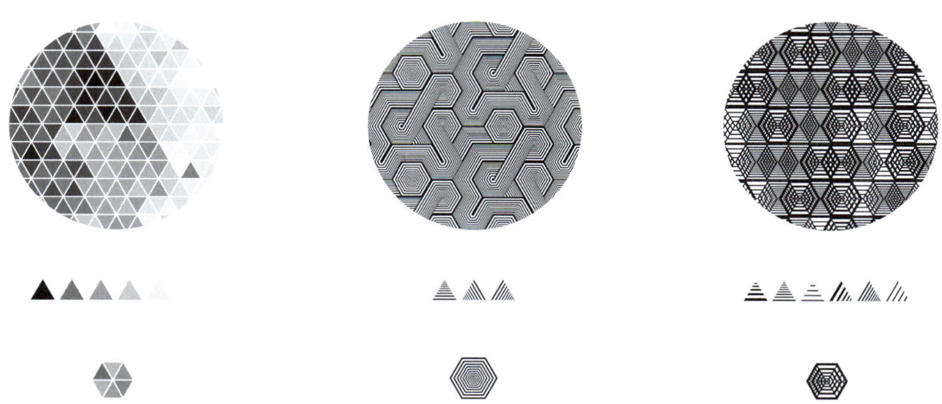

OWNLUCK

Ownluck is a Company dedicated to enhance the experience on transport fluids.

Termluck is a thermo made from metal or plastic, with three divisions that have the capacity to store 333 ml of liquid in each of these.

The icon is a composition of O + L, represented among the 3 divisions. Product packaging is intended to be vibrant and colourful to the public, creating a perfect balance with the clean product. We created a language based on the divisions of the logo, combined with two color palettes.

Design Agency:
Sabbath

Coffee, Tea or Juice ?
3 Divisions of 333 ml.
Study or work,
will never be warmer.

LA NOTE

Paper Tigre is a young stationery brand whose first collection of books and objects. It was founded by five thirty that combine their creative universe to make again the writings valuable. It uses only recycled paper made in France. Today Paper Tigre develops and launches into a range of cosmetics for women. She called the Note, which refers to the note of a fragrance and note that you can register, referring to his many books. The packaging is inspired by his paper world that is influenced by origami.

Designer:
Mathilde Laffon, Julie Ferrieux, Marine Giraud

Client:
N/A, school project

Photography:
Mathilde Laffon, Julie Ferrieux, Marine Giraud

Others:
School: Intuit.lab

la note
EAU DE PARFUM

la note
EAU DE PARFUM

LA NOTE PAR PAPIER TIGRE
L'essence d'une femme libre et audacieuse
Une note féminine avec un caractère affirmé
et pourtant d'une étonnante fraicheur
La note, première fragrance
signée Papier Tigre.
Fabriqué en France
avec joie et bonne humeur.

CRÈME DE JOUR
la note
CRÈME
DE JOUR
la note
EAU DE PARFUM
la note
LOTION
LOTION

PAPIER TIGRE
Paris Tokyo

la note
Le premier parfum
signé Papier Tigre
la note
EAU DE PARFUM

la note
EAU DE PARFUM

la note
EAU DE PARFUM

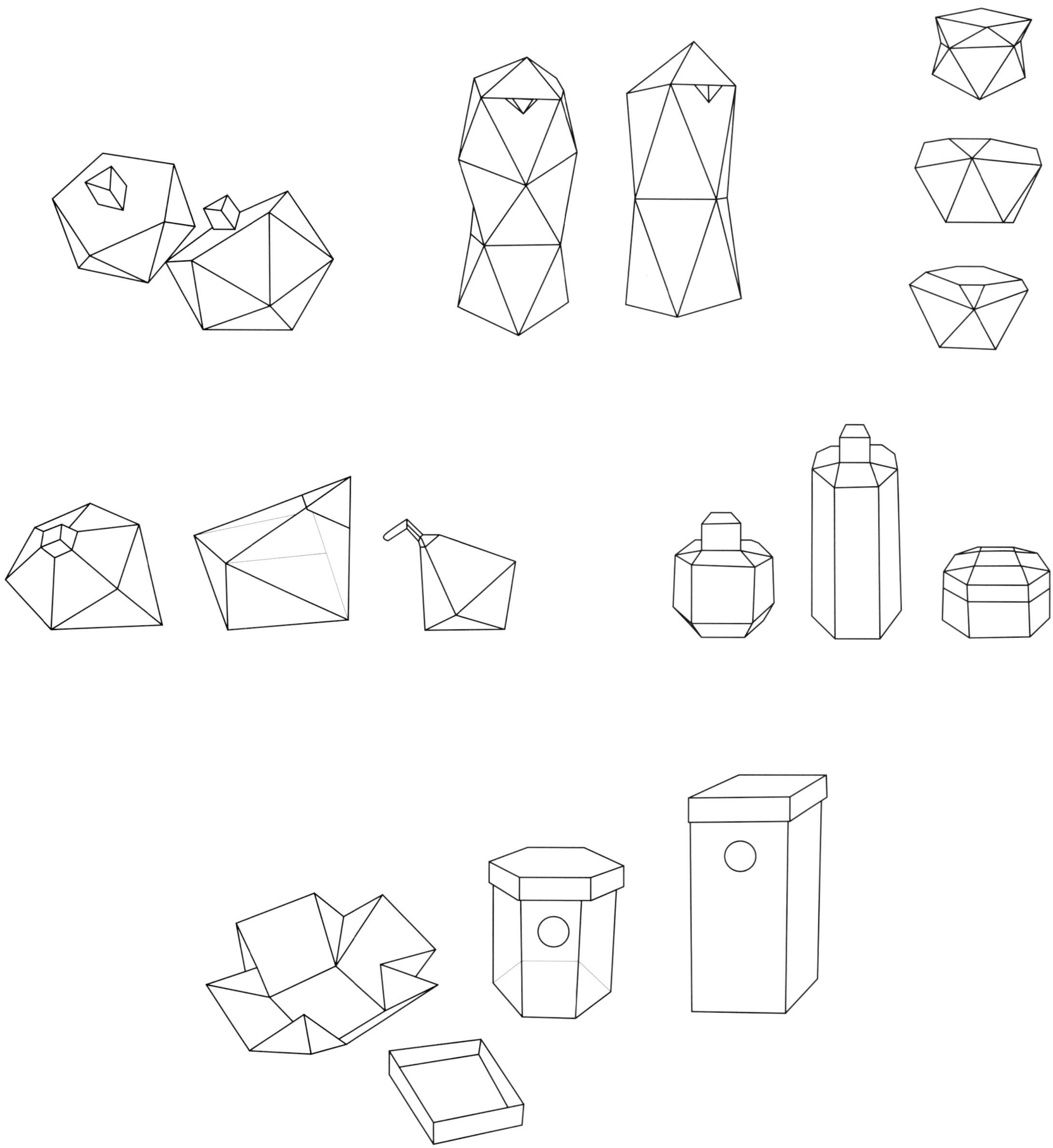

SMART HEADS

The Brief:

Since 1901 the smart heads from all disciplines at Hansgrohe have been revolutionising the world of showers with the latest technology, intuitive operation, maximum sustainability and excellent design.

Hansgrohe presented the smartest showerhead ever:

the Raindance Select 150 EcoSmart.

The task was to develop a high-impact method of drawing the attention of interior designers and bathroom fitters to this innovative product on the bathroom fixtures market using a small budget.

Design Agency:
Kolle Rebbe

Client:
Hansgrohe

Photography:
Frank Stöckel

The idea:

We drew immediate attention to how incredibly smart this shower handset is with a packaging first.

We gave the Raindance Select 150 smart showerheads a face. The face of other smart heads of our time - visionaries, masters and inventors.

Real Smart Heads.

The Solution:

A limited special edition was produced and sent to selected business partners abroad. Impressed calls and emails from around the world were the result.

Based on the enthusiastic response, the decision was then taken to produce Smart Heads for the consumer market.

HANSGROHE = DUSCHVERGNÜGEN².
Diese Theorie lässt sich ganz einfach beweisen. Mit den intelligenten Köpfen von Hansgrohe. Aber Vorsicht: Es geht um die Strahlung. Denn mit dem integrierten Select-Knopf, können Sie gleich zwischen mehreren Strahlarten wechseln. Von weich, voluminös bis kräftig massierend. Bei diesem Duschvergnügen werden Sie die Relativität der Zeit live in Ihrem Badezimmer erleben. Starten Sie gleich selbst den Versuch und Sie werden automatisch die zweite Gleichung verstehen: Einmal Hansgrohe = immer Hansgrohe!
RAIN
RAIN AIR
WHIRL
hansgrohe
SMART HEADS
- DIE INTELLIGENTEN KÖPFE VON HANSGROHE -
SELECT
SELECT

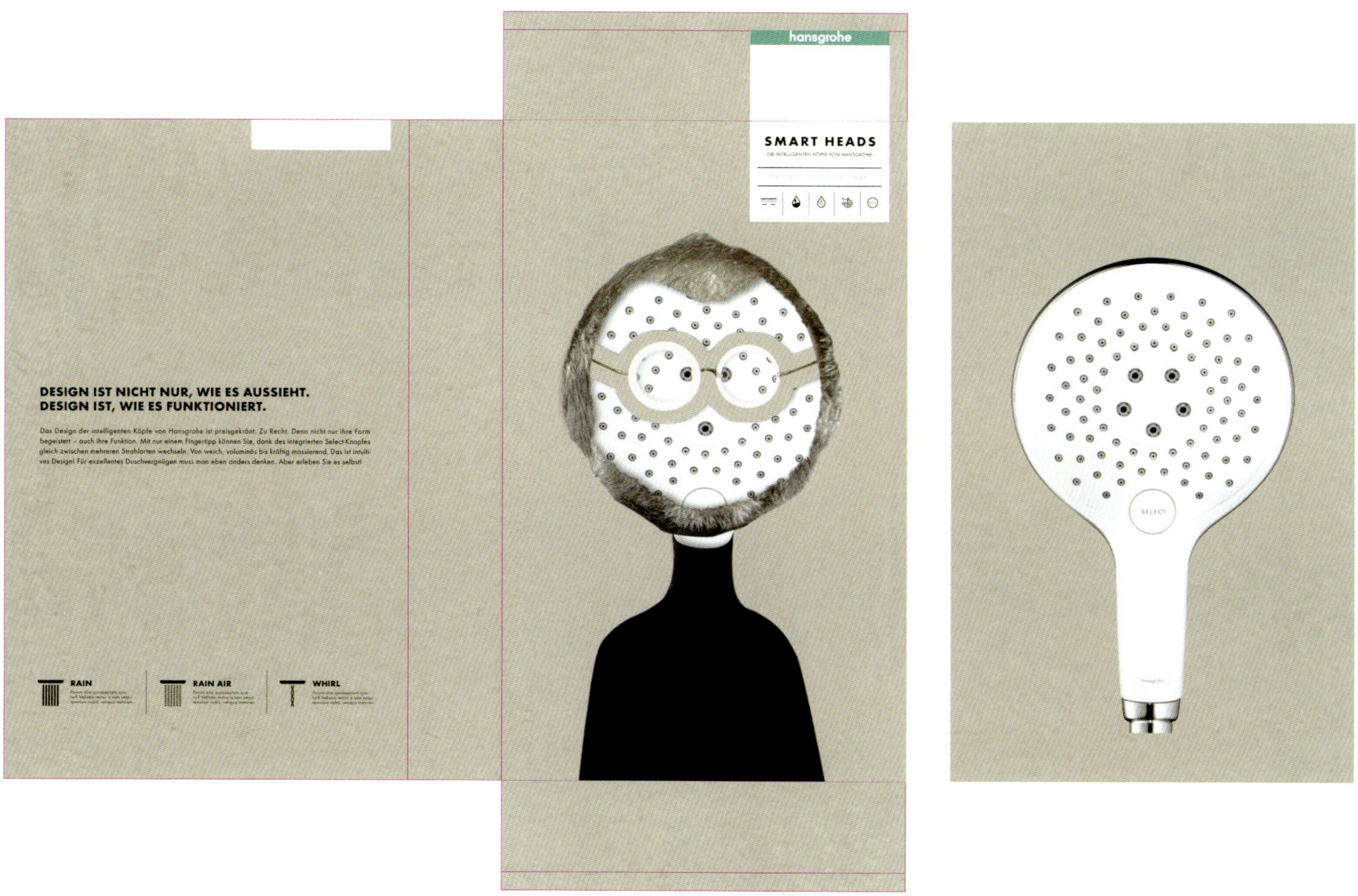
DESIGN IST NICHT NUR, WIE ES AUSSIEHT.
DESIGN IST, WIE ES FUNKTIONIERT.
Das Design der intelligenten Köpfe von Hansgrohe ist preisgekrönt. Zu Recht. Denn nicht nur ihre Form begeistert – auch ihre Funktion. Mit nur einem Fingertipp können Sie, dank des integrierten Select-Knopfes gleich zwischen mehreren Strahlarten wechseln. Von weich, voluminös bis kräftig massierend. Das ist intuitives Design! Für exzellentes Duschvergnügen muss man eben anders denken. Aber erleben Sie es selbst!
RAIN
RAIN AIR
WHIRL
hansgrohe
SMART HEADS
- DIE INTELLIGENTEN KÖPFE VON HANSGROHE -
SELECT

TAIWAN HIGH MOUNTAIN TEA

This series of carefully selected High Mountain tea leaves was harvested from Taiwan's Lishan, Hsi-tou, Alishan, Mt, Yu Ling, and Chilai mountain ranges. Different aromas emanate due to the diversity of the altitudes of the mountains which correspond to maple, bamboo, cedar, plum, pine, and lily plants and also represent six mountain birds; the Varied Tit, Liocichla steerii, Collared Bush Robin, Vinaceous Rosefinch, Strong-footed Bush Warbler and the White-whiskered Laughingthrush. Such widely different natural ecologic aspects instill Taiwan's High Mountain tea with individual characteristics, fantastically giving this distinctive tea a special aftertaste.

The packaging design is based on each mountain's altitude with its corresponding plant and bird to combine artistic conception and skill in traditional ink paintings and modern color ink paintings. The natural ink paintings not only showcase the images of plants and birds which represent each mountain specifically, but they also build an atmosphere of a mountain in the clouds, allowing users to feel as if they are drinking tea in an elegant, peaceful poem or painting.

Design Agency:
Victor Branding Design Crop.

Designer:
Wan Ju Lee

Illustrator:
Qing Ya Yang

Client:
PinShiang Tea Co., LTD.

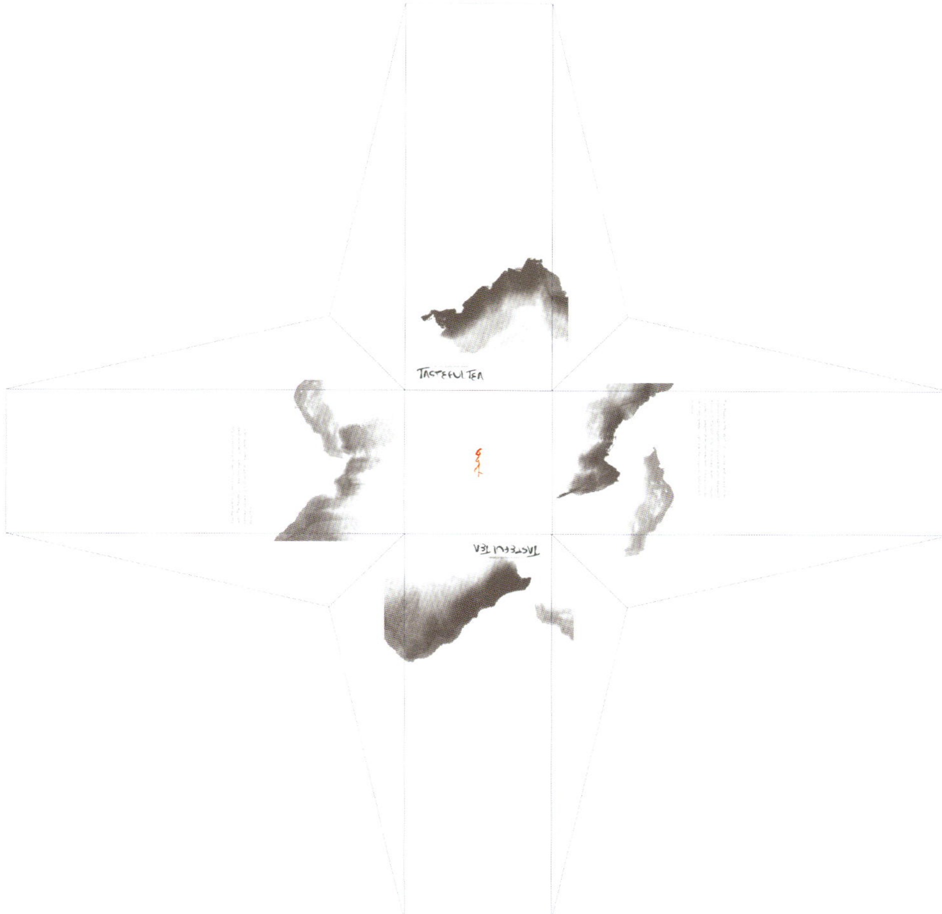
TASTEFUL TEA
TASTEFUL TEA

HIGH MOUNTAIN TEA

SHANNON & LYKKE TEA PACKAGING

Shannon & Lykke Tea Packaging was made by a designer who really appreciates drinking the tea. Without the tea shelf context surrounding this line art packaging, the consumer may not immediately guess the contents of the box. The typical carton for teabags and loose leaf infusions has floral and foliage symbols pictured on it, as well as the word 'tea' printed quite legibly. This design differs in its singular use of abstract symbolism, in its quirky typesetting and in its rather more minimalist and contemporary layout.

On a base of white space, Shannon & Lykke is represented with the name of the brand in bold. What also catches the eye immediately is the artistic motif to symbolize each flavor, printed with great detail in a limited color scheme within each theme. Designer Iryna Korshak played with strokes and geometric shapes to create her highly modern line art packaging that would set such a product apart from competitors.

Designer:
Iryna Korshak

Client:
Shannon&Lykke

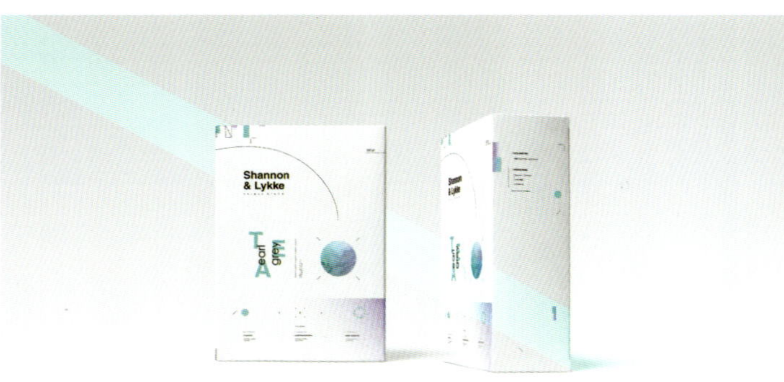

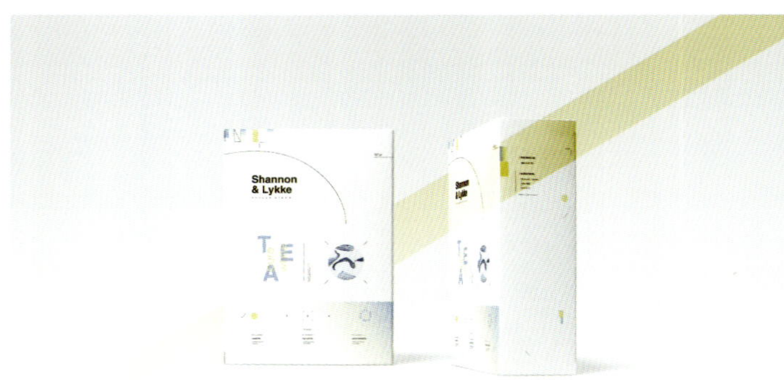

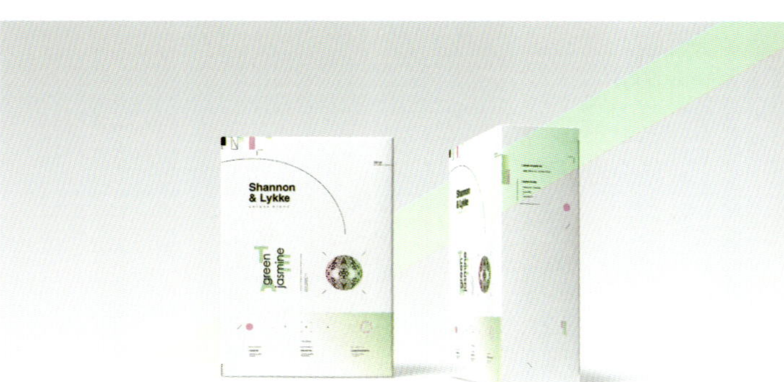

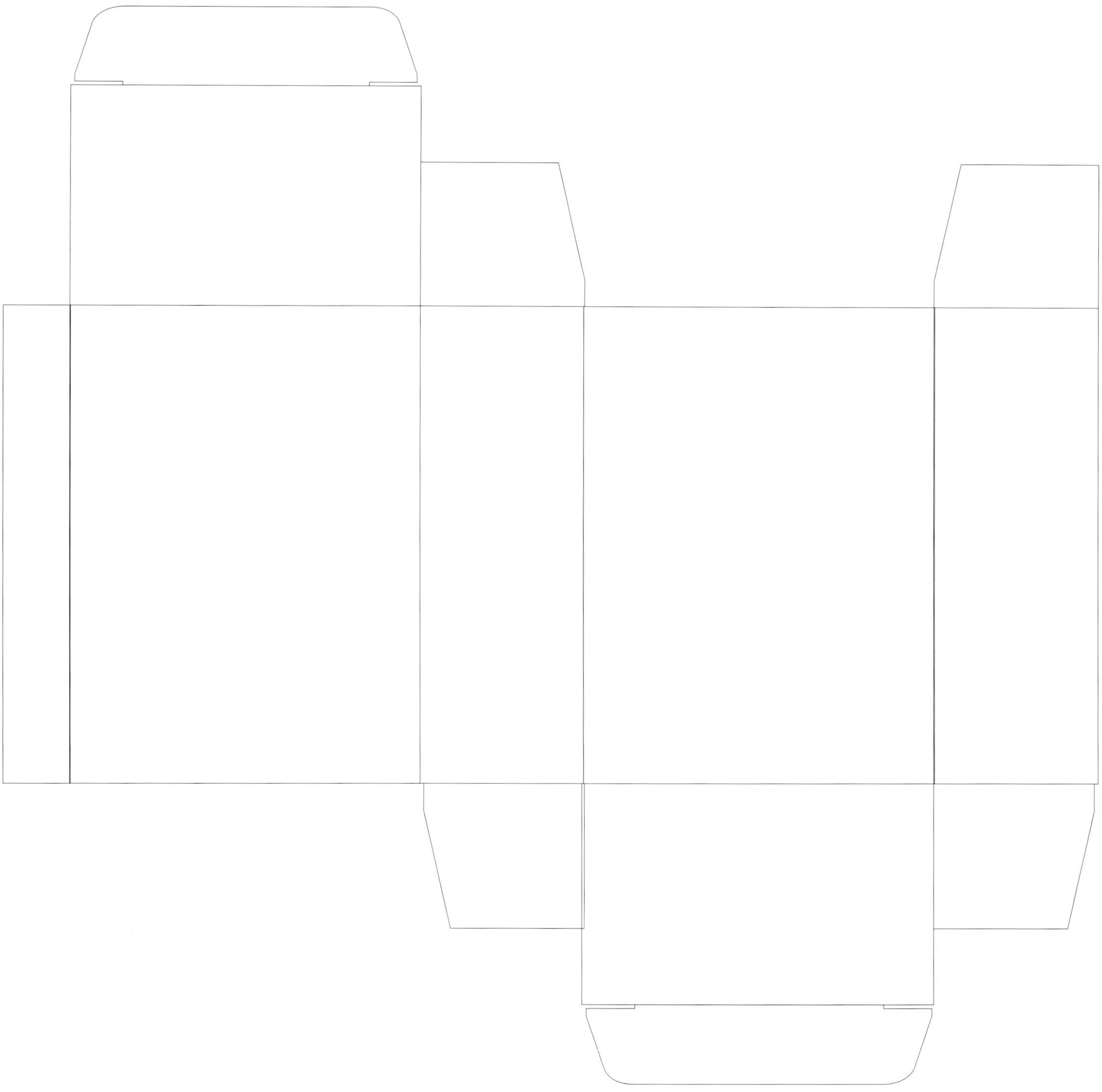

Design Agency:
STUDIO FLAG

Creative Director:
Cheolhee Hwang

Designer:
Cheolhee Hwang

Client:
Sansuhwa Teahouse

It is our intention in this work to symbolize the blending of Korean Hangul typography with splendid scenery of the mountain and river which has expressed in Sansuhwa.

Also, we have used to extend some of these symbols with harmony of straight lines and curves into patterns for various packaging design.

SANSUHWA TEAHOUSE IDENTITY & PACKAGE DESIGN

"Sansuhwa" is premium tea house, located in the heart of Seoul, Hannamdong, with meaning that "good tea comes from pure water and great mountains."

In fact 'Sansuhwa' is a unique genre of Korean traditional landscape painting which portrays the essence of Korean mountains and streams.

The word, San-su, literally means mountains and water in Korean.

BACKHOUSE // SHOT GLASS PACKAGING

My approach was to create a Wild West themed shot glass packaging inspired by the shapes of revolver barrels. I created an imaginary story about about a guy named "Bill Backhouse" who was notorious for defeating his challengers by a drinking duel instead of a gun fighting back in the Wild West. I generated a protective and re-usable packaging structure that transforms into a roulette table for the drinking game based on this hypothetical story. Also, I created a label for an exclusive whiskey inspired by the one Bill Backhouse loved.

Designer:
Yuya Yoshida

School:
School of Graphic design at the Academy of Art University in San Francsico

Instructor:
Thomas McNulty

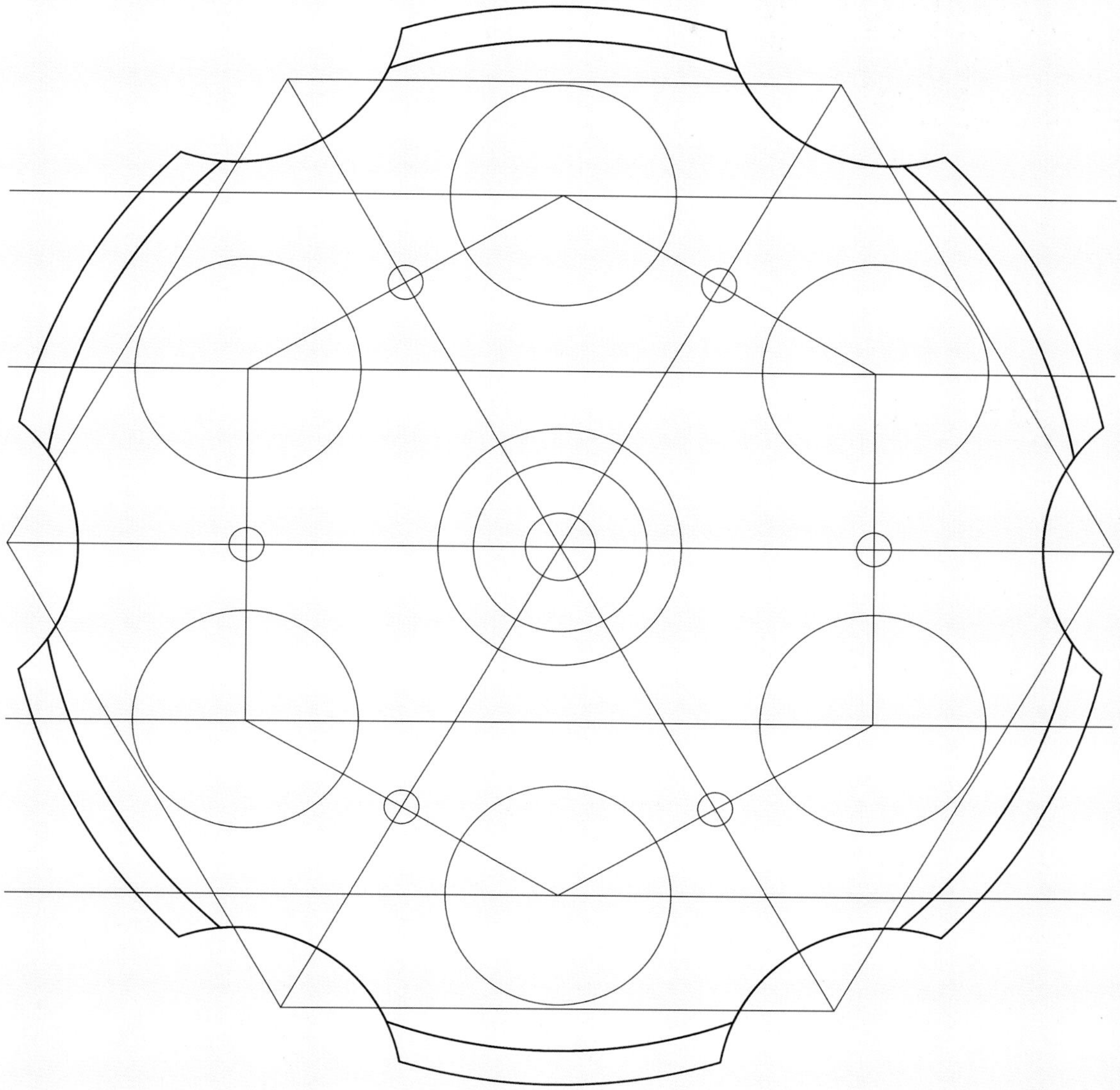

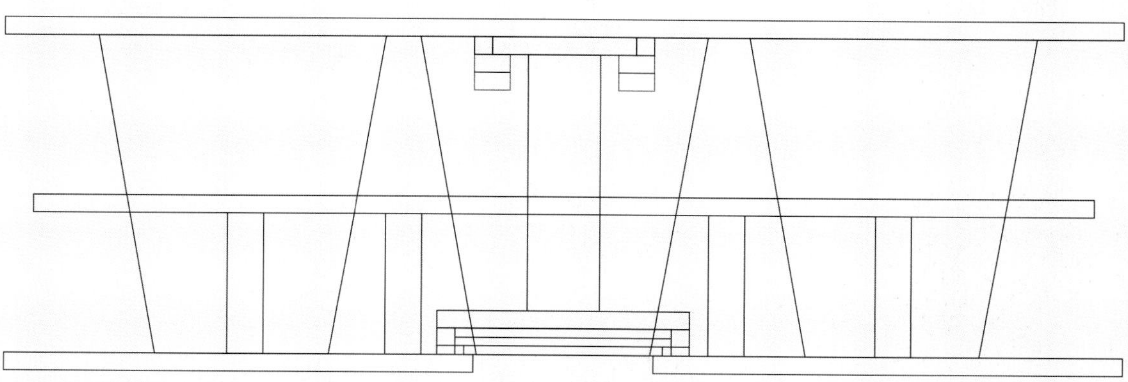

This Hi-End model has a spotless high-class style and sound. Magister can be with equal success used both at home and in a studio. The model is equipped only with the best materials and components. Acoustic enclosure is made of solid Bog oak over 2000 years old. Three years of research and experiments resulted in unique geometry of the acoustic enclosure, which knows no equals in the world.

Design Agency:
Kennerton Audio Equipment

Designer:
Ustinov Kirill

MAGISTER

The name Kennerton, being a combination of words Kenner (Ger. craftsmasters, experts) and Tone (sound, sounding, tone) and a bat-shaped emblem, which, as is well-known, has an exceptionally keen hearing, reflects the brand background and its main aims and purposes pitch-perfectly. This very word describes the company entire history. And it is hardly surprising, since the initial goal of the company was the production of upmarket audio equipment, expensive, luxurious, beautiful and unconditionally high-class in sound.

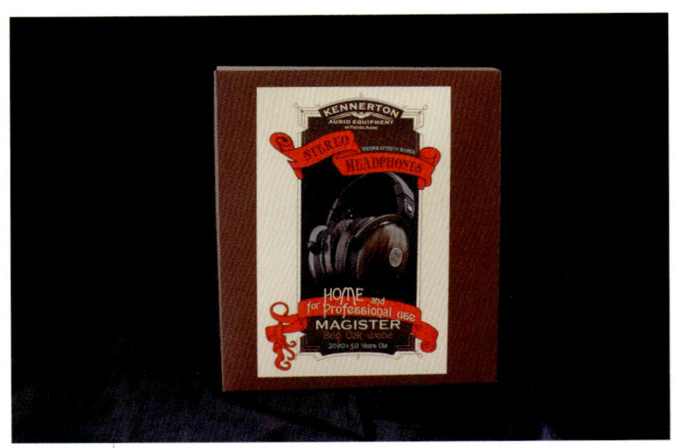

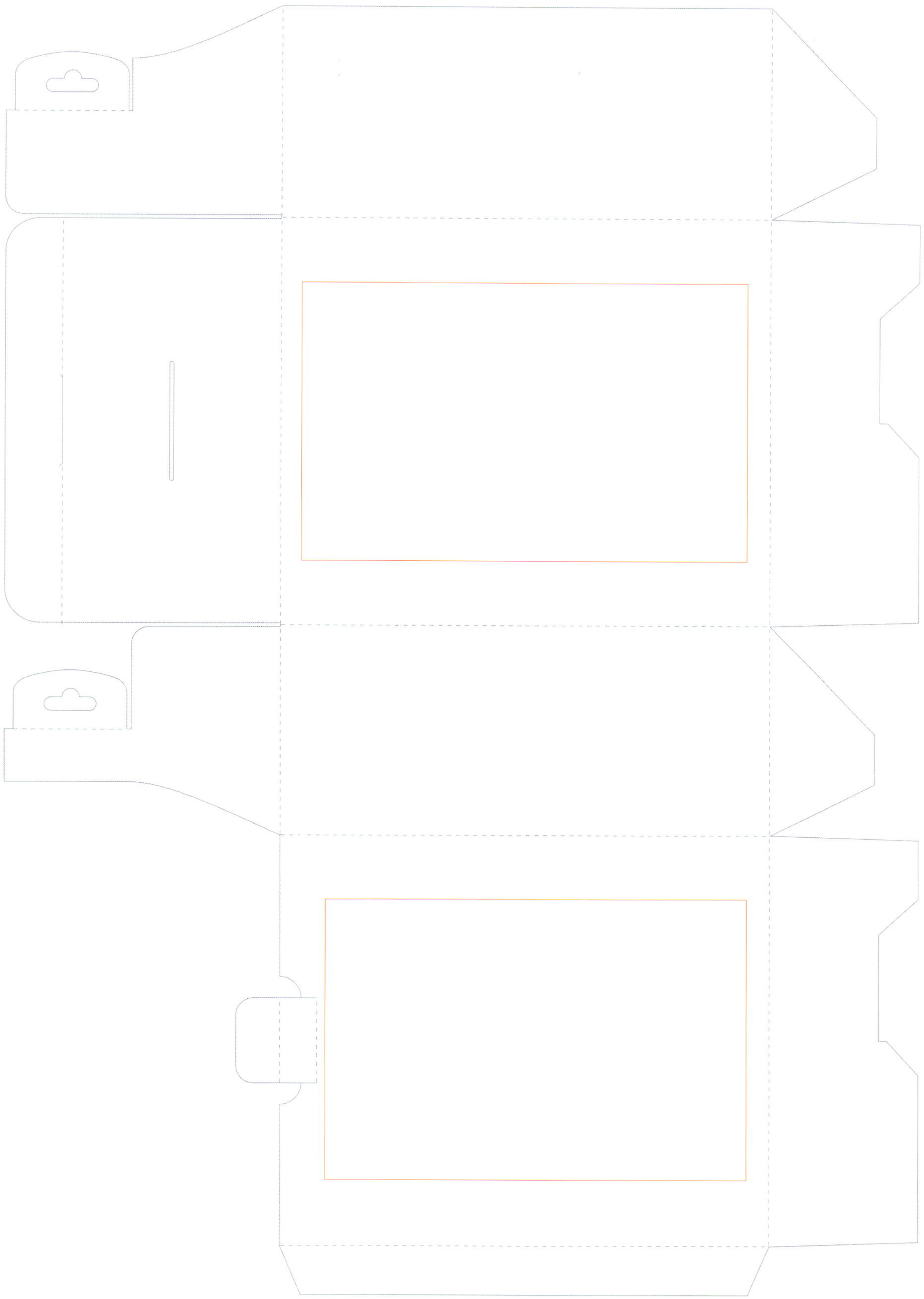

DRINGCUP™ PORCELAIN CUP FAMILY

DringCup is a playful set of porcelain coffee cups designed and made by Imre Eszter, that provides the user the possibility to customize the object themselves. The rings around the cups create a safe distance from the hot surface as well as ensure a comfortable grip instead of the traditional handles. On the front side of the label the logotype can be seen, the necessary information and the description take place on the two outer sides. These sides are fillable, what is a surface for self expression by choosing colour and size, so the package reflects to the playful intention of the cups.

The package allows wrapping both sizes of cups for transport and post. The different sizes have some identical parameters but most of them are different. To prevent moving of the cup inside of the package I designed handles which are fold back into the box. When you put the cup in the package, these handles holds the cup around the rings. The single face corrugated cardboard gives appropriate flexibility, but it is a strong material too. The package is die-cut, flat wise but easy to assemble, sticking free and eco friendly.

Artist:
Molnar Rebeka

Client:
Imre Eszter

Photography:
Richard Kelemen

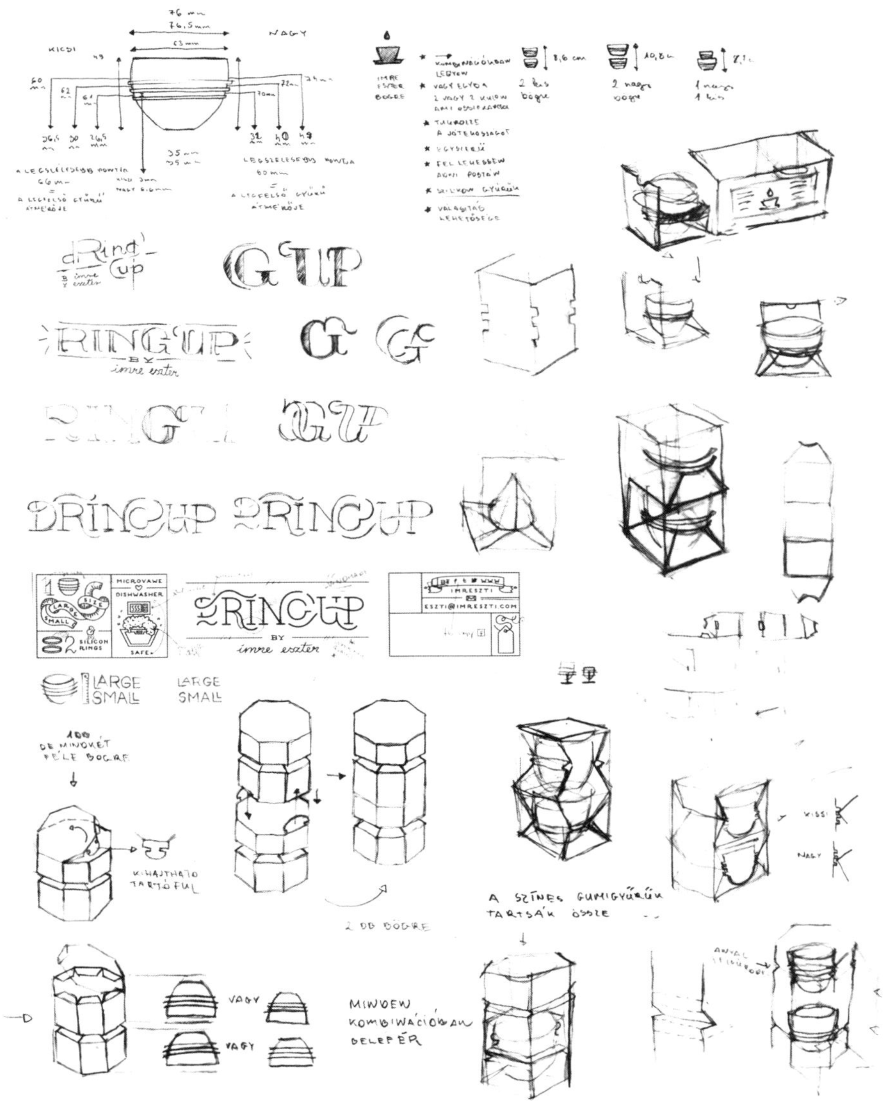
RINGCUP
BY
imre eszter
MICROWAVE
DISHWASHER
SAFE
SILICON RINGS
LARGE
SMALL
VAGY
A SZÍNES GUMIGYŰRŰK TARTSÁK ÖSSZE
MINDEN KOMBINÁCIÓBAN BELEFÉR

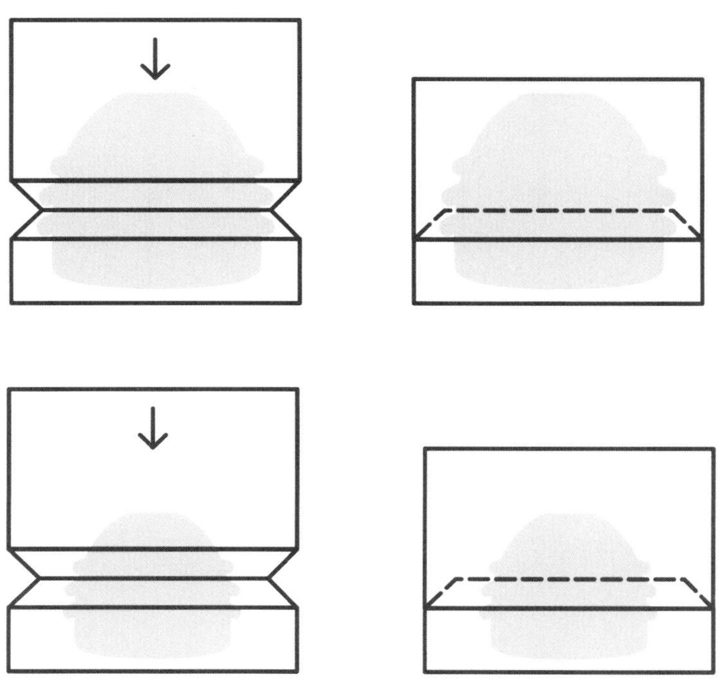

OLDSKOOL 33 1/3

Fischer Audio headphones were developed with the help of state of the art technology, framed in retro-styled housing, they feature natural wood with precise engraving. Their sound takes the best from both era in rich fusion of modern detailed and dynamic sound complemented by natural, full and rich vintage feel. All this is packed in light and portable design, so the headphones just disappear on your head.

Design Agency:
Fischer Audio Engineering

Idea:
Dolgushin Alexey

Designer:
Ustinov Kirill

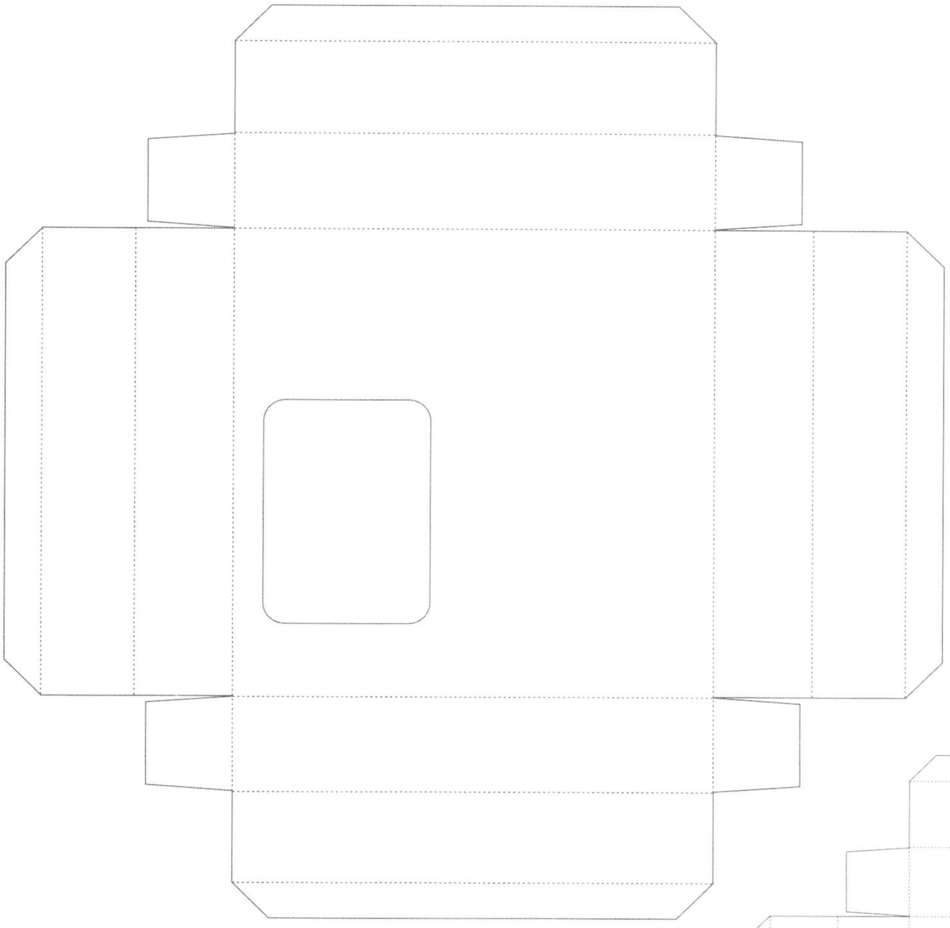

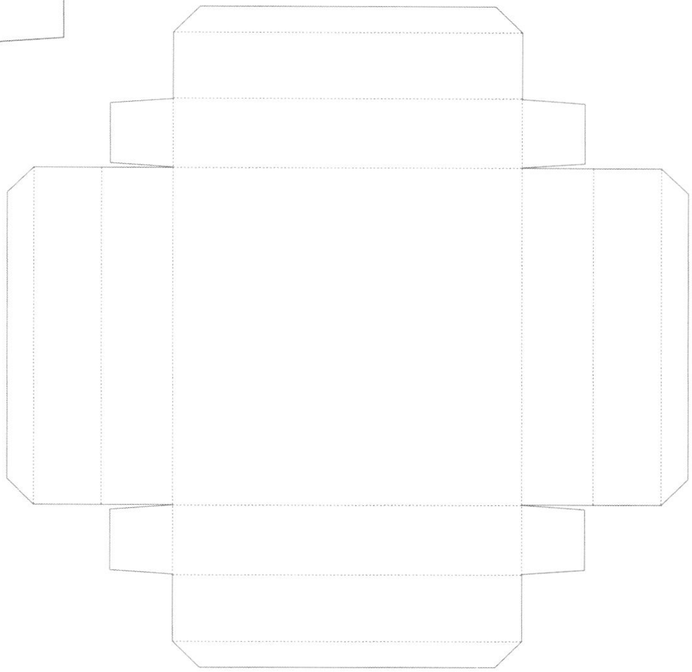

VIZIT

Design Agency:
Vozduh Advertising Agency

Creative Head:
Valery Melnik

Designer:
Roman Luzanov

Design of packagings for TM VIZIT condoms.

Features of each product in a ruler are illustrated by characteristic drawing on a tie, as one of the main man's accessories. The developed visual concept successfully correlates with the name TM.

Color
VIZIT®
12 цветных ароматизированных презервативов
RU ДВЕНАДЦАТЬ ЦВЕТНЫХ АРОМАТИЗИРОВАННЫХ ПРЕЗЕРВАТИВОВ СТАНДАРТНОГО РАЗМЕРА ИЗ ЛАТЕКСА: гладкие, цветные, цилиндрические, с накопителем, с ароматизированной (мята, банан, клубника) силиконовой смазкой.
Номинальные размеры: длина – 180 мм, ширина – 52 мм, толщина стенки – 0,06 мм.
Презервативы при правильном использовании являются эффективным средством для предохранения от нежелательной беременности и заболеваний, передающихся половым путем. Презервативы VIZIT® (ВИЗИТ) изготовлены из натурального латекса и прошли проверку электроникой.
Соответствуют требованиям ГОСТ 4645-81 и ISO 4074:2002. РУ № ФСЗ 2010/07354 от 07.07.2010 г.
⚠ Следуйте прилагаемой инструкции по применению. Срок годности 5 лет.
Презервативы произведены/Презервативи виготовлені/Мішеќаптарды ґндіруші: CPR Produktions- und Vertriebs GmbH, Im Kirchenfelde 8, 31157 Sarstedt, Germany. ЦПР Продуктионс- унд Вертрибс ГмбХ, Им Кирхенфельде 8, 31157 Зарштадт, Германия
Контроль качества и упаковка произведены: ООО «МастерПак», Россия, 141640, Московская обл., Клинский р-н, д. Новощапово, д. 16, стр. 2
Импортер в России: ООО «БОЛЕАР МЕДИКА», юр. и факт. адрес: 115201, г. Москва, Каширский пр-д, д. 13, тел.: +7 (495) 787-92-68, www.bolear.ru
Импортер в Республике Беларусь: ООО «БОЛЕАР Белмед», 220124, г. Минск, ул. М. Лынькова, д. 123-68 (1 этаж), тел.: +375 (17) 205-83-20
UA Презервативи латексні VIZIT Color кольорові ароматизировані. Дванадцять кольорових ароматизованих презервативів стандартного розміру з латексу: гладкі, кольорові, циліндричні, з накопичувачем, з ароматизованою (м'ята, банан, полуниця) силіконовою змазкою.
Номінальні розміри: довжина – 180 мм, ширина – 52 мм, товщина стінки – 0,06 мм.
Презервативи VIZIT (ВІЗІТ) виготовлені з натурального латексу та пройшли перевірку електронікою. Відповідають вимогам ISO 4074:2002.
⚠ Дотримуйтесь інструкції по застосуванню, що додається! Термін придатності 5 років.
Імпортер в Україні: ТОВ «БОЛЕАР МЕД», 04073, м. Київ, вул. Куренівська, 18, тел. +38 (044) 531-39-46
CP 9763/2010 03.09.2010
KZ Он екі тїрлі-тїсті латекстен жасалынєан стандартты мґлшерімен мішеќаптар: тегіс, тїрлі-тїсті, цилиндр тїрізді, жинаєышы бар, хош иістендірілген (жалбыз, банан, ќўлпынай иісті) силиконды майлаєышымен.
Номиналды мґлшері: ўзындыєы – 180 мм, ені – 52 мм, ќабырєа жуандыєы – 0,06 мм.
VIZIT (ВИЗИТ) мішеќаптары табиєи латекстен жасалєан жјне электрондыќ баќылаудан ґткен.
МЕМСТ 4645-81 жјне ISO 4074:2002 талаптарєа сай болады.
Ќоса берілген нўсќаулыќты басшылыќќа алыѕыз. Жарамдылыќ мерзімі 5 жыл.
Ќазаќстандаєы импорттаушы: «Болеар Казмед» ЖШС, 050008, Алматы ќ-сы, Мысбаев к-сі, 43 їй, тел.: +7 (727) 379-90-79
DE Zwölf bunte Kondoms mit Aroma in Standard-Größe aus Latex: glatt, bunt, zylindrisch, mit Speicherplatz, mit Aromasilikonfeuchtbeschichtung (Minze, Banane, Erdbeere).
Maße: Länge – 180 mm, Breite – 52 mm, Dicke – 0,06 mm.
Jeder Kondom einzeln elektronisch getestet.
Entsprechen den Normen ISO 4074:2002.
Richten Sie sich nach der beigelegten Vorschrift. Haltbarkeitsdauer: 5 Jahre.
Hersteller der Rohkondome: CPR Produktions- und Vertriebs GmbH, Im Kirchenfelde 8, 31157 Sarstedt, Germany.
Importeur in Russland: OOO «BOLEAR MEDIKA»,Kaschirskij projesd, 13, 115201, Moskau, tel.: +7 (495) 787-92-68, www.bolear.ru
www.vizit-condoms.ru
Color
VIZIT®
12 цветных ароматизированных презервативов
количество -12 шт.
VIZIT®
AR46
25°C
0°C
Color
VIZIT®
12 цветных ароматизированных презервативов
4 601834 004262
Годен до
Використати до
Дейін жарамды
LOT
Партия №
Партія №
VER 1

KINKY TOYS

So all this makes it a low entry product, for young exploring adults or for everyone who's into trying these products, but never had the courage to actually get one. Usually you need to go to specialized shops for these matters, where the packaging is mostly vulgar, offensive, and therefore less popular to place in your shopping bags. For most adults, walking into a the shop, is already a big deal, so they never reach to these products. By creating a packaging like this, with a soft approach, these products can easily be sold in more common stores, which open doors to a wider range of consumers.

Design Agency:
QUATRE MAINS, Belgium

Designer:
Patrick De Grande, Kobe De Keyser, Hendrik Colenbier

CAT´S PLAY HOUSE

The project begins with customer demand for a solution of promotional packaging for 4 packages of pounds to cat and a package of 10 pounds of litter (7 kg in total).

The solution presented to the client is a conventional box, but thanks to a series of pre - cuts that transforms and generates an additional product for promoting the transformation of the initial cash in a home for the cat. The success of this approach is possible because the instinct of cats hiding indoors, deeste mode is the opportunity to supplement the packing finding a common interest to make, cat owner and cat. This design opens up new opportunities to create an extension of the life of the packaging (principle of sustainability or life cycle) and above extension of the brand presence in the buyer (extended relationship Product and service), additionally it creates a new element in the promotion and creation of more extensive and lasting ties with the brand, beyond the consumption of the food. Graphically has a clean image but still respecting the brand icons generating connection between users.

Design Agency:
VAYA! Agency

Designer:
I.D. Guil lermo Acevedo Beltrán

Client:
MARS of Colombia

Author of All the Graphic Material:
I.D. Guillermo Acevedo Beltrán

whiskas
whiskas

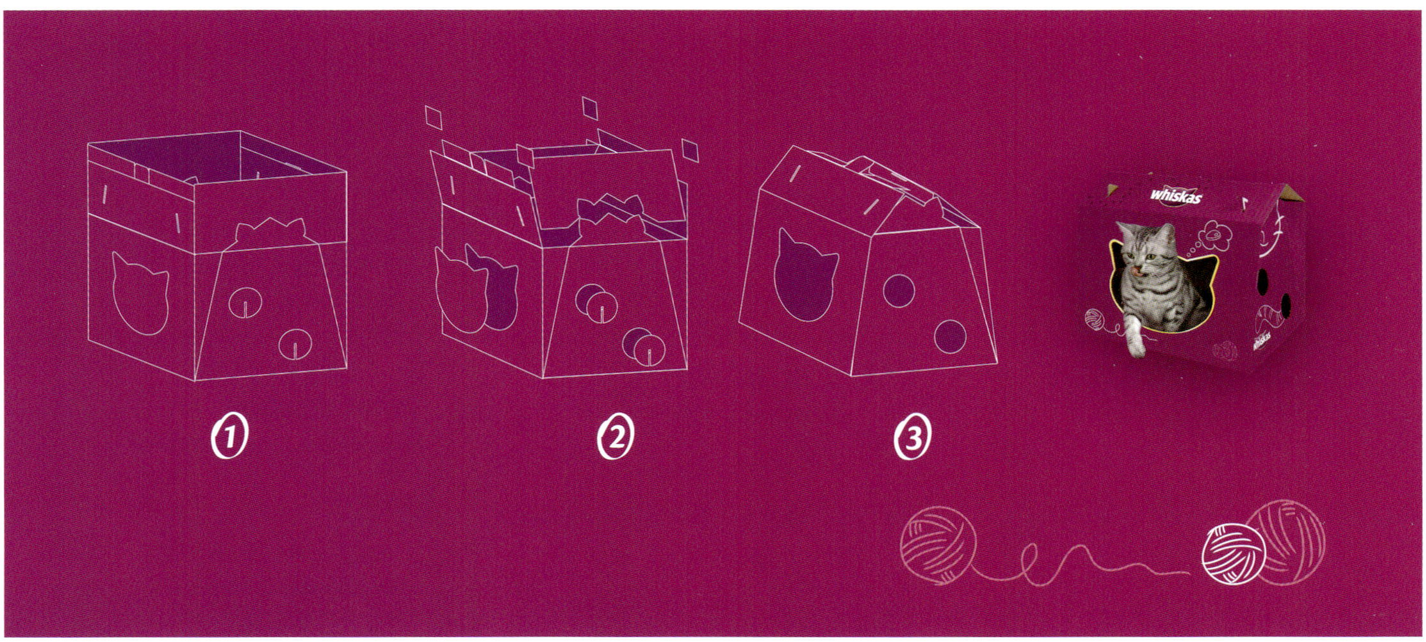
1
2
3
whiskas

CONTRIBUTORS

5.5 DESIGNSTUDIO
www.5-5designstudio.com
P210

ADRIANNA BRONISZEWSKA
www.behance.net/AdaBroniszewska
P070

AEKYUNG DESIGN CENTER
www.aekyung.kr
P022

AFTERHOURS
www.afterhoursdesign.co.uk
P220

AIZAWA OFFICE
www.aizawa-office.com
P068

ALESSIA SISTORI
cargocollective.com/alessias
P088

ALEXANDRE MENDES
www.behance.net/AlexMendes
P208

ALICE MACAROVA
www.behance.net/macarova
P094

ANDREY KLEMENKOV
cocha-bepeck.com
P078

ANGELINI DESIGN
www.angelinidesign.com
P230

ATTILA ÁCS
www.behance.net/attilaacs
P238

BIG FISH
www.bigfish.co.uk
P080

BINALOGUE
binalogue.com
P160

BRANDIZIAC BRANDING AGENCY
www.brandiziac.ru
P156

BRIELLE WILSON
www.briellewilson.com
P232

CARLOS NOGUEIRA
carlosnogueira.com.br
P126

CATHERINE MARION
marioncatherine.prosite.com
P058

CHAD WITZEL
cwitzel.com
P172

CHRISTIAN PANNICKE
www.christianpannicke.de
P074

DAPY
www.dapyparis.com
P192

DEREK DUBLER
derekdubler.com
P086

DIEGO FRAYLE
www.diegofrayle.co
P054, P062

EGGPLANT FACTORY
www.eggplantfactory.co.kr
P096

ESTABLISHED
www.establishednyc.com
P016, P018, P026, P030

ESZTER VARGA
www.behance.net/esztervarga
P102

EVE WARREN
evewarren.com
P098

FABIAN FOHRER
www.fabianfohrer.com
P176

FISCHER AUDIO
www.fischeraudio.com/index.php/en
P256, P260

FOXTROT STUDIO
www.foxtrotstudio.pl
P186

GEOMETRY GLOBAL
www.geometry.com
P052, P224

GUILLERMO ACEVEDO BELTRÁN | VAYA! AGENCY
www.vayagencia.com
P266

GUSTAV KARLSSON
www.gustavkarlsson.com
P092

HANNAH HART
www.hannahhartdesign.com
P180

INORAMA ILLUSTRATION STUDIO
inorama.ru
P112

ION LUCIN
www.behance.net/ionL
P178

IRYNA KORSHAK
www.behance.net/IraKorshak
P250

JAKOB + MACFARLANE
www.jakobmacfarlane.com
P168

KARIM RASHID
www.karimrashid.com
P012, P014, P108, P174

KOLLE REBBE
www.kolle-rebbe.de
P246

KOREFE
korefe.de
P024

LADOLEA
www.ladolea.gr
P064

LA FONDA GRÀFICA
www.lafondagrafica.com
P194

LINDA HURTIG
www.lindahurtig.com
P042

LOGIC DESIGN
www.logic-design.eu
P164

MARA HOLTERDORF
www.behance.net/MaraHolterdorf
P088

MARIA DURIANA RODRIGUEZ
www.behance.net/DurianaRodriguez
P054, P062, P200

MARIA MORDVINTSEVA-KEELER
www.behance.net/mord
P128

MARIA PONOMAREVA
www.behance.net/ma_nia
P162, P216

MARKO DANILOVIĆ
www.nordeus.com
P150

MATHILDE LAFFON
mathildelaffon.com
P242

MILLFORD
millford.nl
P228

MILK BRANDING PROFESSIONALS
www.milk.com.gr
P196, P198

MOUNTAIN
www.mountain.nl
P134, P136, P182

MOLNAR REBEKA
www.behance.net/RebekaMolnar
P258

OKTA DESIGN
oktadesign.com.br
P082

PATRICK VEILLET
www.patrickveillet.com
P028, P032, P034

PEDRO PAULINO
pedropaulino.com
P138

P&W
www.p-and-w.com
P142, P144

QUATRE MAINS
www.quatremains.be
P106, P264

RENAN VIZZOTTO
www.behance.net/ARTURENAN
P202

REVERSE INNOVATION
www.reverseinnovation.com
P046, P084, P204

RÚBEN RODRIGUES
rubro-del-rubro.tumblr.com
P110

SABBATH
www.sabbathvisuals.com
P240

SILVIA ALBERTÍ
www.behance.net/SilviaAlbertiColomar
P062

SOTANO STUDIO
www.sotanostudio.com
P008

SOLOVYOVDESIGN
www.solovyovdesign.by
P146

SPICE DESIGN
www.spice.com.br
P158

SPRINGETTS BRAND DESIGN CONSULTANTS
www.springetts.co.uk
P148

STAS NERETIN
www.behance.net/1iz25
P038

STEVE SIMPSON
www.stevesimpson.com
P190

STUDIO AIO
studioaio.com
P114

STUDIO FLAG
www.designbyflag.com
P252

SVOE MNENIE
www.svoemnenie.com
P212

TANGRAM DESIGN
www.tangram.no
P120

VICTOR BRANDING DESIGN CROP.
www.victad.com.tw
P044, P118, P124, P132, P206, P234, P248

VOZDUH ADVERTISING AGENCY
www.ra-vozduh.ru
P048, P262

YUYA YOSHIDA
yuyayoshida.com
P254

ARTPOWER

Acknowledgements

We would like to thank all the designers and companies who made significant contributions to the compilation of this book. Without them, this project would not have been possible. We would also like to thank many others whose names did not appear on the credits, but made specific input and support for the project from beginning to end.

Future Editions

If you would like to contribute to the next edition of Artpower, please email us your details to: artpower@artpower.com.cn